Wave Propagation in Materials for Modern Applications

Wave Propagation in Materials for Modern Applications

Editor

Carlos Granger

scitus
academics

Wave Propagation in Materials for Modern Applications
Edited by **Carlos Granger**

ISBN: 978-1-68117-609-3
Library of Congress Control Number: 2016946464

616, Corporate Way, Suite 2, 4766,
Valley Cottage, NY 10989

www.scitusacademics.com

Notice

Printed in United States of America on Acid Free Paper

Preface

There is an increasing interest in micro and nanotechnology in the present time. Nanotechnology offers advances in new types and new applications of materials with unique electromagnetic and physical characteristics. The communication process involves the transmission of information from one location to the next. As we have seen, modulation is used to encoding the information on a carrier wave with analog or perhaps electronic methods. It is only the carrier wave characteristics that determine exactly how the signal propagates over a significant distance. Propagation of wave is one of the ways waves travel. As regards the path of oscillation in relation to the direction of propagation, we can distinguish between longitudinal wave and transverse wave. Propagation in a vacuum and in a material medium can occur in electromagnetic waves. A neighborhood disturbance in the magnetic and electrical fields produces an electromagnetic wave. From its origin, the wave propagates outwards in all directions. If the place where it propagates (for example, air) is the same everywhere, the trend will spread uniformly in every direction.

Wave Propagation in Materials for Modern Applications is dedicated to modern acoustic and electrodynamics methods designed to describe wave propagation in these contemporary materials and nanodevices.

Table Of Contents

CHAPTER 1

Electromagnetic Waves Absorbing Characteristics of Composite Material Containing Carbonyl Iron Particles

Victor A. Zhuravlev, Valentin I. Suslyaev, Evgeny Yu. Korovin, Kyrill V. Dorozhkin

Department of Radioelectronics, National Research Tomsk State University, Tomsk, Russia

ABSTRACT

Results of measurements of permeability, permittivity and radar absorption properties of composites on basis of carbonyl iron particles R-10 brand are presented in this paper. The calculations and experimental studies have shown that in the super high frequency (SHF) and extremely high frequency (EHF) ranges on the basis of two-layer structures with different content of carbonyl iron particles can create a radar absorbing coatings with a reflectivity of less than −10 dB over a wide bandwidth from 3.1 to 17.1 GHz and from 27 to 37 GHz. Absorbing properties of composites are saved in terahertz frequency range from 250 to 525 GHz.

KEYWORDS

Electromagnetic Wave Absorbers, Radar Absorbing Materials and Coatings, Composite Materials, Carbonyl Iron Powders, Spectrum of Permittivity and Permeability

1. INTRODUCTION

Active development of the microwave frequency range sharpened the need for magnetic materials absorbing electromagnetic radiation. Such materials are required to reduce harmful influence on biological objects, to provide electromagnetic compatibility of units and blocks of high-frequency devices, to build anechoic chambers and to protect information. Composite radar absorbing coating (RAC) which consists of a matrix of high-mole- cular polymer compounds and filler particles of ferromagnetic or ferrimagnetic materials are widely used [1]. Such materials are most effectively used as electromagnetic waves absorbers (EMWA) located on the metal surfaces.

The carbonyl iron powders (CIP) are used at works [2] -[10] as fillers for composite RAC in microwave range. Investigation of complex permeability ($\mu = \mu' - i\mu''$) and permittivity ($\varepsilon = \varepsilon' - i\varepsilon''$) spectra of composites with different concentration of CIP at frequencies up to 18 GHz shows that these materials have noticeable permeability and permittivity loses, and they are perspective as EMWA for microwave frequency range. The rather narrow operating frequency band is disadvantage of such RAC.

It is shown in [5] [6] that double-layer RAC is more broadband and has less reflection coefficient than single- layer RAC [1] -[3] [7] -[10] . The first layer of such RAC is an absorbing layer. It is located on the metal surface. The second layer is matching with free space. It has smaller values of the permeability and permittivity.

It is known [11] that frequency band of high magnetic loses caused by natural ferromagnetic resonance at presence of domain structure lies generally in the limits from $2\pi f_1 = \gamma H_a$ to $2\pi f_2 = \gamma(H_a + 4\pi M_S)$. Because of big value of saturation magnetization of iron $M_S \approx 1.7$ kGs the frequency band of high magnetic loses is from $f_1 \approx 1.4$ GHz up to $f_2 \approx 61$ GHz. In calculation it is accepted $\gamma/2\pi = 2.8$ GHz/kOe, anisotropy field $H_a \approx 0.5$ kOe. This estimation shows that composite materials on basis of CIP can have noticeable absorbing properties in wide frequency range, including SHF and EHF.

Results of investigation of permittivity and permeability spectra, absorbing properties of composite materials composed of CIP of R-10 brand in polyurethane matrix are presented in this paper. CIP of R-10 brand had onion like structure with average size of particles about 3.5 μm. Thickness of investigated flat samples was 1 mm , volume content of CIP 9% (sample 1) and 53% (sample 2).

2. THEORY

Theoretical analysis of radar absorbing properties of materials is usually carried out for the simplest case of normal incidence of a plane electromagnetic wave (EMW) on a magnetodielectric (MD) layer. Permeability (μ) and permittivity (e) of layer are taken as complex scalar quantities. For developing of radio protective materials and coatings located in free space layer of MD without metallization is considered. For reduction of radar signature a model of MD layer disposed on the metal surface is used. Formulas for calculating the complex reflection (R_T) and transmission (T_T) coefficients in the first case can be written in the form [12]:

$$R_T = \frac{\rho(1-\exp(-2i\gamma d))}{1-\rho^2\exp(-2i\gamma d)}, \quad T_T = \frac{(1-\rho^2)\exp(-2i\gamma d)}{1-\rho^2\exp(-2i\gamma d)}. \tag{1}$$

Here: $\rho=(Z-1)/(Z+1)$—reflection coefficient from the front edge of MD. $Z=\sqrt{\mu/\varepsilon}$—wave impedance.

$\gamma=k_0\sqrt{\varepsilon\mu}$—EMW propagation constant in MD. $k_0=\omega/c$—the wave number of free space. $\omega=2\pi f$—cir- cular frequency of the EMW. c—the speed of light. d—thickness of the layer. $i=\sqrt{-1}$.

Expression for the reflection coefficient (R_R) for located on the a metal plate layer of MD has the form [12] :

$$R_R = [\rho-\exp(-2i\gamma d)]/[1-\rho\exp(-2i\gamma d)] \tag{2}$$

Commercial advantage has RAC with a small thickness and weight. Therefore it is interesting to analyze the expression (1) and (2) in the case of a thin layer and find out which material parameters are crucial to improve the efficiency of consumer qualities of such materials. Use the approximation of a electrically thin layer: $|\gamma d| \ll 1$. In the linear approximation formulas (1) are significantly simplified and coefficients R_T, T_T can be rewritten as:

$$T_T \approx [1+i(\mu+\varepsilon)k_0 d/2]^{-1}, \; R_T \approx iT_T(\mu-\varepsilon)k_0 d/2. \tag{3}$$

For the modules of the transmission and reflection coefficients are obtained the following expressions:

$$|T_T| = 1/\sqrt{1+k_0 d(\mu''+\varepsilon'')}, \; |R_T| = |T_T|\frac{k_0 d}{2}\sqrt{(\mu'-\varepsilon')^2+(\mu''-\varepsilon'')^2} \tag{4}$$

According to formulas (3), (4) in the thin layer approximation a module of the transmission coefficients depends only on the imaginary parts of permittivity and permeability describing the magnetic and dielectric losses in the MD layer. The reflection coefficient vanishes when real and imaginary parts of the permeability and permittivity are equal. In this case there is equality of the wave impedances of the interface MD layer and free space. In real composites to achieve such equality it is impossible.

The expression (2) for the module of reflection coefficient located on a metal surface layer of MD at $|\gamma d| \ll 1$ can be written:

$$|R_R| = \sqrt{(1-2k_0 d\mu'')/(1+2k_0 d\mu'')} \tag{5}$$

The reflection coefficient depends only on the imaginary part of permeability MD in this case. The optimal thickness of the layer (d_{opt}) ensuring the absence of the reflected wave is equal to:

$$d_{opt} = 1/2k_0\mu'' = c/2\omega\mu''. \tag{6}$$

Equation (6) contains an important consequence for practical applications. To ensure the broad band RAC the imaginary part of permittivity of MD should vary inversely proportional to the frequency:

$$\mu'' \sim 1/\omega. \tag{7}$$

It should be noted that the formulas (4) and (5) provide an opportunity to solve another important experimental problem—to assess the values of the imaginary components μ'', ε'' from the measured values of $|R_R|$ and $|T_T|$.

Calculation of the reflection coefficient of multilayer RAC was carried out in [13]. The formula for the reflection coefficient from the interface between the n + 1 and n layers is follows:

$$R_{n+1} = \left(Z_{\text{in}}^{(n)} - Z_{n+1}\right) / \left(Z_{\text{in}}^{(n)} + Z_{n+1}\right) \tag{8}$$

Z_{n+1}—impedance of layer n + 1. $Z_{\text{in}}^{(n)}$ —input impedance of layer n. It is equal to

$$Z_{\text{in}}^{(n)} = \frac{1 + R_n \exp(-2i\gamma_n d_n)}{1 - R_n \exp(-2i\gamma_n d_n)} Z_n \tag{9}$$

Equations (8) and (9) will be used later to calculate the reflection coefficient of the two-layer RAC located on the metal surface.

3. RESULTS AND DISCUSSIONS

3.1. Permeability and Permittivity Spectra of Composite

Spectra of permeability ($\mu = \mu' - i\mu''$) and permittivity ($\varepsilon = \varepsilon' - i\varepsilon''$) measured using vector network analyzer Agilent Technologies E8363B. Measurements were performed in the frequency range 10 MHz - 20 GHz. A coaxial waveguide (R/T method, [12]), an irregular microstrip resonator [14] [15] , and a set of rectangular multimode volume resonators [16] were used as measuring cells.

Since various experimental techniques have different systematic and random measurement errors, permeability and permittivity values measured by the corresponding methods at the ends of the examined frequency range differed. In this regard, the experimental magnetic spectra of the composites were processed by the method suggested in [17] [18] . To match the frequency dependences of the imaginary and real permeability components, the Kramers-Kronig relations and independent measurements of the initial static permeability (μ_0) of the materials were used. Here μ_0 values, equal to 10.6 and 3.1 for composites with CIP contents of 53% and 9%, respectively, were measured for toroidal samples at a frequency range 20 Hz to 2 MHz using precision LCR Meter Agilent E4980A.

Figure 1 and Figure 2 shows the magnetic permeability spectra of the composites.

It can be seen that a maximum in the frequency dependence of the imaginary part of permeability for both samples is localized near a frequency of 1 GHz and depends weakly on the filler concentration. The measured real and imaginary permeability components are correlated with the data presented in [7] -[10] for close values of CIP concentrations.

Our measurements of the permittivity spectra demonstrated that its variance was small in the examined frequency range, and the permittivity of samples remains virtually constant: $\varepsilon' = 28$, $\varepsilon'' = 4$ for the sample 2 and $\varepsilon' = 6.5$, $\varepsilon'' = 0.2$ for the sample 1.

3.2. Single-Layer EMWA SHF Range

It was shown above (see Equation (7)) that to develop thin broadband RAC on the metal surface, it is necessary that μ'' varies inversely proportional to the frequency of the EMW. In the case of composite materials based on CIP, the imaginary permeability component decreases with frequency much more slowly.

Figure 3 and Figure 4 shows the frequency dependence of the reflection coefficients of the layers located on the metal surface calculated according to formula (2). The obtained above values of μ and e samples 1 and 2 are used in the calculations. Numbers near the curves-sample thicknesses in mm.

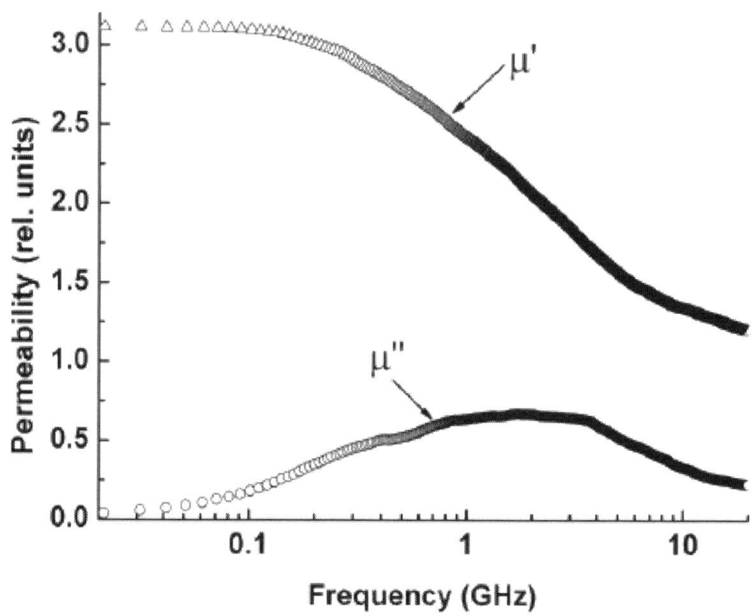

Figure 1. Permeability spectra of the sample 1 with CIP content of 9%.

Presented in Figure 3, Figure 4 results demonstrated that on the base of sample 1 and 2 it is difficult to make a single-layer RAC with the reflection coefficient less than −10 dB for the frequency band exceeding the octave.

3.3. Two-Layer EMWA SHF Range

We investigated a two-layer absorber design whose absorbing layer on metal surface was formed by the composite sample 2 (thick d_1) and whose matching layer was formed by the sample 1 (thick d_2) with small values of the permeability and permittivity. The calculation was performed according to the formulas (8) and (9). The simulation results presented on Figure 5 by solid lines.

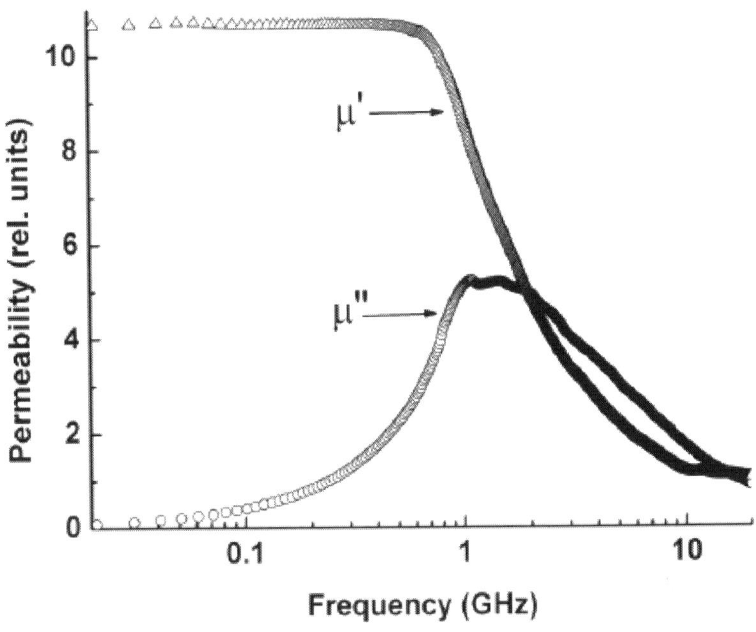

Figure 2. Permeability spectra of the sample 2 with CIP content of 53%.

Curve 1—the calculation for the absorbing layer thickness $d_1 = 0.9$ mm, matching layer $d_2 = 2.3$ mm. At such thicknesses reflection coefficient smaller −10 dB is observed in the frequency range from 3.1 to 17.1 GHz. Decrease in the thickness of the layers to $d_1 = 0.78$ mm and $d_2 = 1.95$ mm leads to a shift of the frequency band with a reflection coefficient ≤−10 dB in the high frequency region from 4 to 20 GHz (curve 2).

The points in Figure 5—experimental curve measured in the coaxial short-circuited line size of 7/3.05 mm. Measurements were performed on a two-layer construction with thickness d_1 = 1 mm (sample 2) and d_2 = 2 mm (sample 1). It can be seen that the deviation from of optimal thickness in the experimental RAC resulted in an increase of the reflection coefficient in the region of frequencies from 6 to 11 GHz. Thus, the use of bilayer RAC leads to substantial expansion of the band of operating frequencies as compared with monolayer.

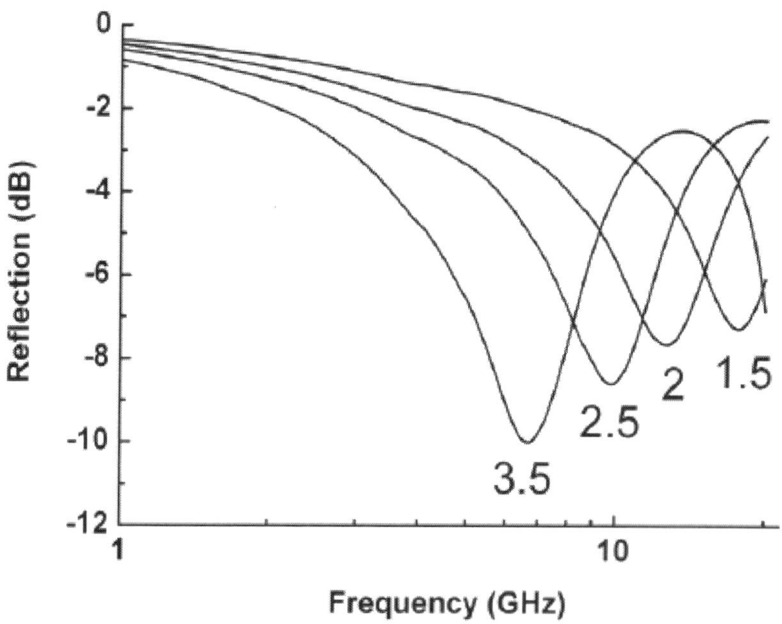

Figure 3. Reflection coefficients of the sample 1 calculated for different sample thicknesses. The numbers near the curves-RAC thickness in mm.

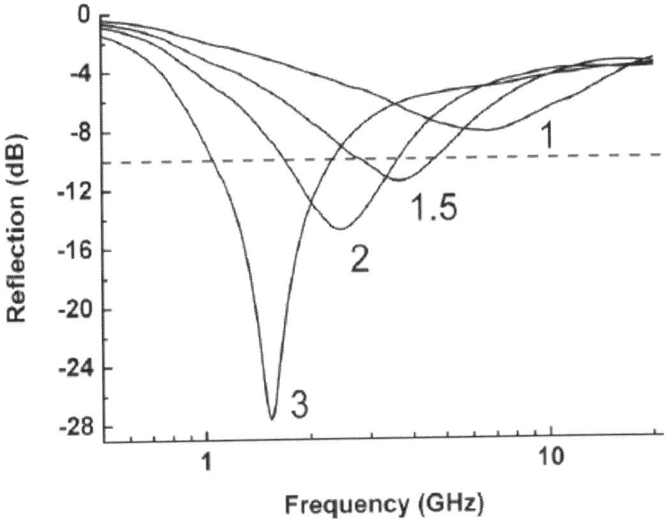

Figure 4. Reflection coefficients of the sample 2 calculated for different sample thicknesses. The numbers near the curves-RAC thickness in mm.

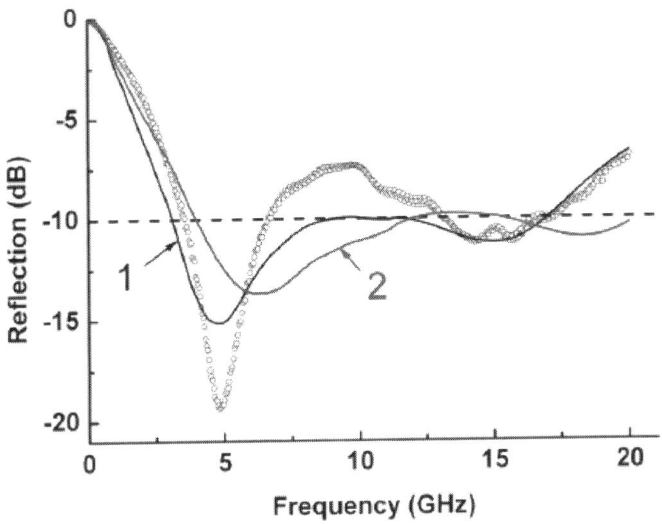

Figure 5. Reflectivity of two-layer RAC. Solid lines—calculation. Curve 1: d_1 = 0.9 mm, d_2 = 2.3 mm. Curve 2: d_1 = 0.78 mm, d_2 = 1.95 mm. Point —an experimental results on a sample with thickness d_1 = 1 mm, d_2= 2 mm.

3.4. Single-Layer and Two-Layer EMWA EHF Range

This section presents the results of experimental investigations of the absorbing properties of composites samples 1 and 2 for frequencies in the range 26 - 37 GHz. Measurements were performed using the rectangular horn antennas 5 cm long and 2 × 1 cm^2 antenna apertures. The SWR of the system of horn antennas without sample in the examined frequency range did not exceed 1.2.

Figure 6 shows the reflection coefficient of RAC located on the metal surface.

Here curve 1 is for sample 1 with thickness of d = 1 mm, curve 2 is for sample 2 (d = 1 mm), and curve 3 is for a two-layer design consisting of sample 1 (d_2 = 1 mm) and sample 2 (d_1 = 1 mm). From the figure it can be seen that the two-layer structure has the reflection coefficient much less than the single-layer structure, and a minimum of reflection is observed at frequencies greater than 37 GHz. As previously, sample 1 with the small content of CIP plays the role of a matching layer.

Figure 7 shows the results of investigations of the electromagnetic wave absorbers in the transmission/ref- lection mode. Values of the transmission and the reflection coefficients in decibels were measured. Then the measured values were recalculated into the transmission coefficient T and the reflection coefficient R on power in percent. The absorption coefficient was calculated from the formula A = 100 − R − T.

Figure 7(a) shows the measured values for sample 1. It can be seen that the reflection coefficient does not exceed 20%, and the transmission and absorption coefficients are close in values (about 40%). The corresponding values measured for sample 2 are shown in Figure 7(b). For this material with large CIP content, the reflection coefficient is somewhat higher. However, the absorbing properties of these RAC are almost twice greater. The transmission coefficient is about 1%. Figure 7(c) shows the measured values for the two-layer RAC design with 1-mm thick sample 1 used as a matching layer and sample 2 also with a thickness of 1 mm used as an absorbing layer. It can be seen that the two-layer absorber of electromagnetic waves in the given range of frequencies has the absorption coefficient greater than 90% and the reflection coefficient smaller than 10%. The transmission coefficient is small; it does not exceed 1%. Thus, good microwave properties of composites based on CIP observed in the EHF range. This is due to the wide frequency band in which there are losses on EFMR.

3.5. Study of the Samples 1 and 2 in the Terahertz Frequency Range

The electromagnetic response of samples 1 and 2 was also investigated in frequency range 150 - 525 GHz. The terahertz spectrometer STD-21 with backward-wave oscillators were used as a source of electromagnetic wave. We investigated the frequency dependence of transmission and reflection coefficients. The results of measuring complex permittivity in this frequency range are shown inFigure 8 and Figure 9.

It can be seen that the real and imaginary parts of permittivity vary slightly within a given range of frequencies. Determined from experiments values of permittivity correlate with the above performed estimates for e in SHF range.

Transmission, reflection and absorption coefficient on power for samples 1 and 2 are shown inFigure 10 and Figure 11.

Figure 10 and Figure 11 shows that the transmission coefficient for both samples is small and does not exceed 1%. For the sample with low content of CIP reflectance in the investigated frequency range does not exceed 10% and that of sample 2, it is about 30%. The remainder part of EMW power is absorbed. Thus, investigated composites have good absorbing properties in the terahertz frequency range.

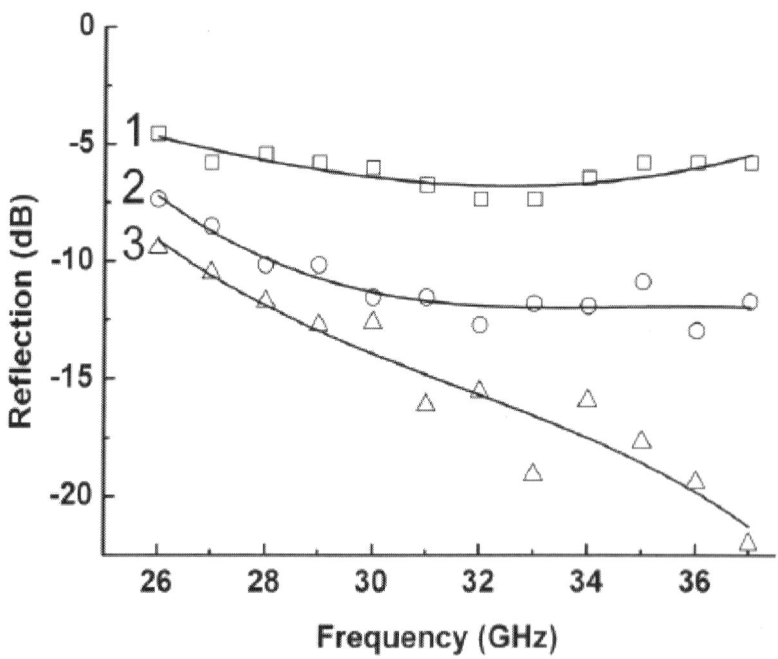

Figure 6. Reflection coefficients from RAC on the metal surface. Curve 1: sample 1 (d = 1 mm). Curve 2: sample 2 (d = 1 mm). Curve 3: two- layer RAC: the sample 1 (d_2 = 1 mm) + sample 2 (d_1 = 1 mm).

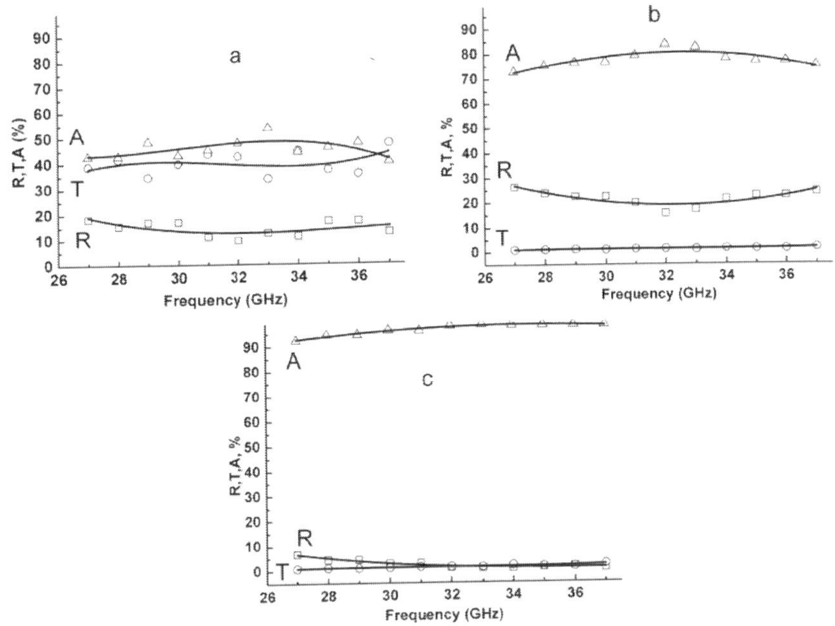

Figure 7. RAC reflection (R), transmission (T), and absorption (A) coefficients. (a) sample 1 (d = 1 mm); (b) sample 2 (d = 1 mm); (c) double layer RAC: the sample 1 (d_2 = 1 mm) + sample 2 (d_1 = 1 mm).

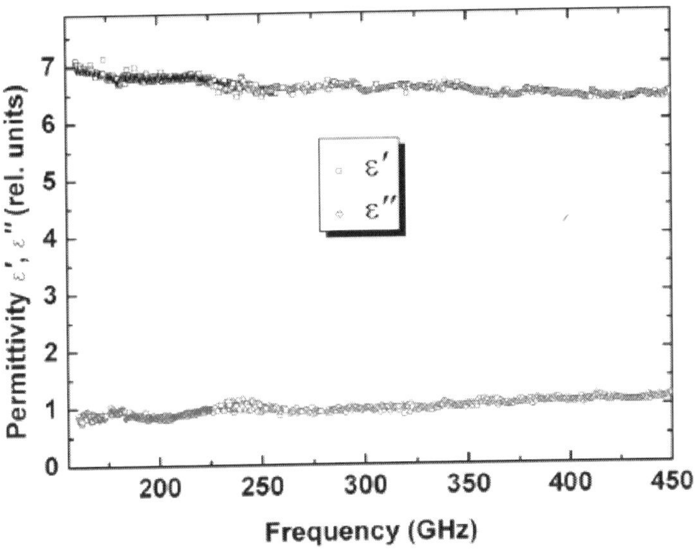

Figure 8. Permittivity spectra of sample 1.

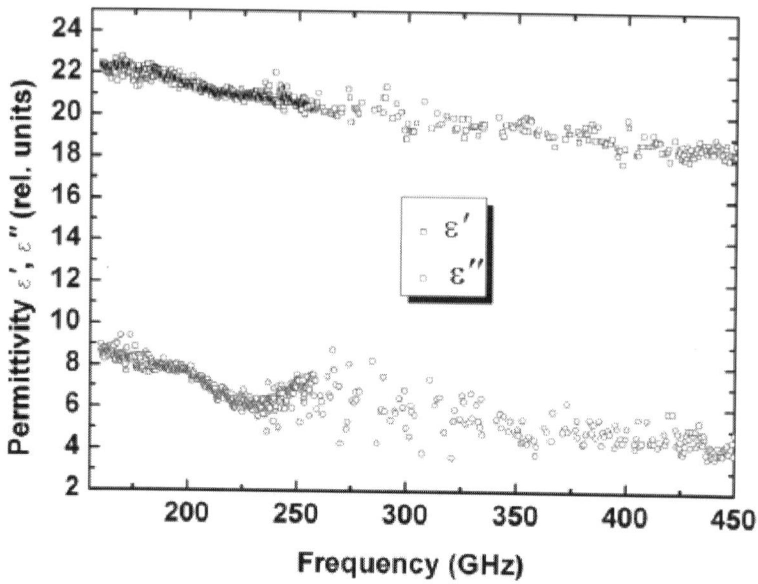

Figure 9. Permittivity spectra of sample 2.

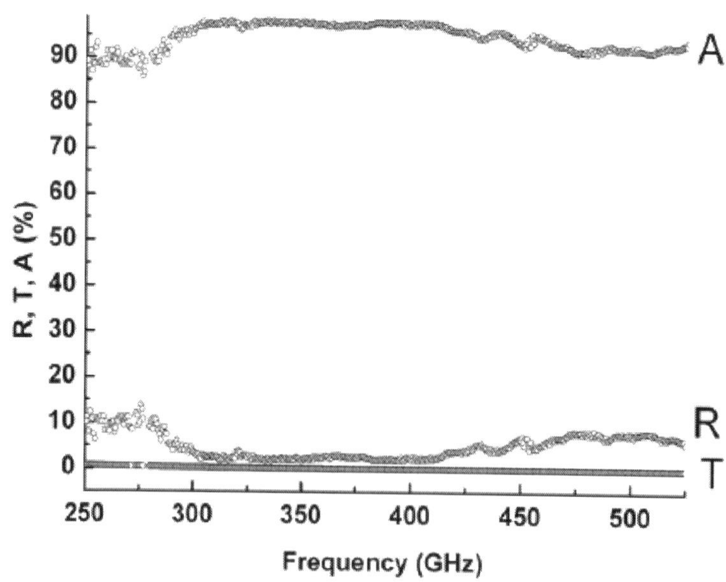

Figure 10. The reflection (R), transmission (T) and absorption (A) coefficients of the sample 1.

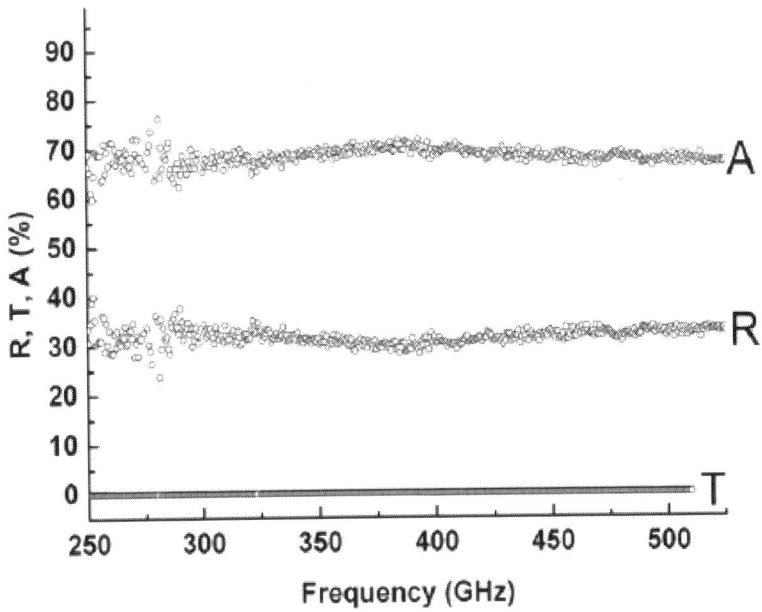

Figure 11. The reflection (R), transmission (T) and absorption (A) coefficients of the sample 2.

4. CONCLUSION

Thus, studies of the permeability and permittivity spectra of the two samples of composites containing CIP in SHF range are carried out. Studies of different designs EMWA have shown that two-layer composite materials based on CIP with different filling factor allow selective absorption with a low reflectance in the SHF and EHF frequency bands. They can be used both as a masking coatings disposed on a metal surface, and as a protective nonreflecting coating with small reflection and transmission coefficients. Sample 1 exhibits good RAC properties also in the frequency range from 250 to 525 GHz.

ACKNOWLEDGEMENTS

Authors would like to thank center of Tomsk Regional Common Use Center: Center of radio-physics measurements, diagnostic and researching of parameters of natural and artificial materials (e-mail:susl@mail.tsu.ru) [19].

Victor A. Zhuravlev, Valentin I. Suslyaev, Evgeny Yu. Korovin, Kyrill V. Dorozhkin This work was supported by RFBR grant No. 13-02-12240.

REFERENCES

1. Bystrov, R.P., Dmitriev, V.G., Zemskii, Yu.A., Perunov, Yu.M. and Cherepenin, V.A. (2012) Special Features of the Development of the Radio-Technical Systems of the Radio-Electronic Struggle. Uspekhi Sovremennoi Radioelektroniki (Achievements of Modern Radioelectronics), 8, 3-28.

2. Wang, M., Duan, Y.P., Liu, S.H., Li, X.G. and Ji, Z.J. (2009) Absorption Properties of Carbonyl-Iron/Carbon Black Double-Layer Microwave Absorbers. Journal of Magnetism and Magnetic Materials, 321, 3442-3446.

3. Liu, L., Duan, Y., Liu, S., Chen, L. and Guo, J. (2010) Microwave Absorption Properties of One Thin Sheet Employing Carbonyl-Iron Powder and Chlorinated Polyethylene. Journal of Magnetism and Magnetic Materials, 322, 1736- 1740.

4. Cheng, Y.L., Dai, J.M., Wu, D.J. and Sun, Y.P. (2010) Electromagnetic and Microwave Absorption Properties of Carbonyl Iron/$La_{0.6}Sr_{0.4}MnO_3$ Composites. Journal of Magnetism and Magnetic Materials, 322, 97-101.

5. Zhuravlev, V.A., Suslyaev V.I. and Korovin, E.Yu. (2010) Dynamic Magnetic Characteristics of a Composite Polymeric Material Based on Carbonyl Iron. Russian Physics Journal, 53, 541-543.

6. Zhuravlev, V.A., Suslyaev V.I., Dotsenko, O.A. and Babinovich, A.N. (2011) Composite Radio-Absorbing Material Based on Carbonyl Iron for Millimeter Wavelength Range. Russian Physics Journal, 53, 874-876.

7. Tan, Y., Tang, J., Deng, A., Wu, Q., Zhang, T. and Li, H. (2013) Magnetic Properties and Microwave Absorption Properties of Chlorosulfonated Polyethylene Matrices Containing Graphite and Carbonyl-Iron Powder. Journal of Magnetism and Magnetic Materials, 326, 41-44.

8. Xu, Y., Zhang, D., Cai, J., Yuan, L. and Zhang, W. (2013) Microwave Absorbing Property of Silicone Rubber Composites with Added Carbonyl Iron Particles and Graphite Platelet. Journal of Magnetism and Magnetic Materials, 327, 82-86.

9. Ding, Q., Zhang, M., Zhang, C. and Qian, T. (2013) Synthesis and Absorbing Mechanism of Two-Layer Microwave Absorbers Containing Polycrystalline Iron Fibers and Carbonyl Iron. Journal of Magnetism and Magnetic Materials, 331, 77-81.

10. Zhang, W., Bie, S., Chen, H., Lu, Y. and Jiang, J. (2014) Electromagnetic and Microwave Absorption Properties of Carbonyl Iron/MnO_2 Composite. Journal of Magnetism and Magnetic Materials, 358-359, 1-4.

11. Gurevich, A.G. and Melkov, G.A. (1996) Magnetization Oscillations and Waves. CRC Press, Boca Raton.

12. Baker-Jarvis, J., Janezic, M.D., Grosvenor Jr., J.H. and Geyer, R.G. (1993) Transmission/Reflection and Short-Circuit Line

Methods for Measuring Permittivity and Permeability. NIST Technical Note 1355.

13. Brekhovskikh, L.M. (1973) Waves in Layered Media. Nauka.

14. Belyaev, B.A., Zhuravlev, V.A., Kirichenko, V.I., Suslyaev, V.I. and Tyurnev, V.V. (1992) Use of an Irregular Microstrip Resonator to Investigate Microwave Properties of Dielectrics with Broad Conductivity Ranges. Measurement Techniques, 35, 992-994.

15. Suslyaev, V.I., Kochetkova, T.D., Zhuravlev, V.A. and Sudakov, S.V. (2003) An Automated Setup for Investigating the Temperature Dependence of the Permittivity Spectra of Polar Liquids in the Range of 0.1 - 1.25 GHz. Instruments and Experimental Techniques, 46, 672-676.

16. Naiden, E.P., Suslyaev, V.I., Bir, A.V. and Politov, M.V. (2004) Permeability Spectra Hexaferrite Nanopowders. Journal of Structural Chemistry, 45, 102-105.

17. Zhuravlev, V.A. and Suslyaev, V.I. (2006) Analysis of the Microwave Magnetic Permeability Spectra of Ferrites with Hexagonal Structure. Russian Physics Journal, 49, 1032-1037.

18. Zhuravlev, V.A. and Suslyaev, V.I. (2006) Analysis and Correction of the Magnetic Permeability Spectra of $Ba_3Co_2Fe_{24}O_{41}$ Hexaferrite by Using Cramers-Kronig Relations. Russian Physics Journal, 49, 840-846.

19. The Electronic Resource. http://www.ckp.tsu.ru/ckp_3/

CHAPTER 2

Computational Study of the Effect of Cortical Porosity on Ultrasound Wave Propagation in Healthy and Osteoporotic Long Bones

Vassiliki T. Potsika [1], Konstantinos N. Grivas [2], Theodoros Gortsas [2], Gianluca Iori [3], Vasilios C. Protopappas [1], Kay Raum [3], Demosthenes Polyzos [2] and Dimitrios I. Fotiadis [1,4,*]

[1] Unit of Medical Technology and Intelligent Information Systems, Department of Materials Science and Engineering, University of Ioannina, GR 45110 Ioannina, Greece

[2] Department of Mechanical Engineering and Aeronautics, University of Patras, GR 26500 Patras, Greece

[3] Berlin-Brandenburg School for Regenerative Therapies, Charité-Universitätsmedizin Berlin, Augustenburger Platz 1, 13353 Berlin, Germany

[4] Foundation for Research and Technology–Hellas, Institute of Molecular Biology and Biotechnology, Department of Biomedical Research, GR 45110 Ioannina, Greece

ABSTRACT

Computational studies on the evaluation of bone status in cases of pathologies have gained significant interest in recent years. This work presents a parametric and systematic numerical study on ultrasound propagation in cortical bone models to investigate the effect of changes in cortical porosity and the occurrence of large basic multicellular units, simply called non-refilled resorption lacunae (RL), on the velocity of the first arriving signal (FAS). Two-dimensional geometries of cortical bone are established for various microstructural models mimicking normal and pathological tissue states. Emphasis is given on the detection of RL formation which may provoke the thinning of the cortical cortex and the increase of porosity at a later stage of the disease. The central excitation frequencies 0.5 and 1 MHz are examined. The proposed configuration consists of one point source and multiple successive receivers in order to calculate the FAS velocity in small propagation paths (local velocity) and derive a variation profile along the cortical surface. It was shown that: (a) the local FAS velocity can capture porosity changes including the occurrence of RL with different number, size and depth of formation; and (b) the excitation frequency 0.5 MHz is more sensitive for the assessment of cortical microstructure.

KEYWORDS

ultrasound; bone modeling; osteoporosis; cortical porosity; basic multicellular units

1. INTRODUCTION

Osteoporosis is a bone pathology characterized by low bone mass and structural deterioration of bone tissue leading to bone fragility, thinning of the cortical cortex, reduced bone strength and an increased susceptibility to fractures. Nowadays, the ageing of the population has led to greater prevalence of osteoporosis which is responsible for approximately 2 million fractures which occur annually and may lead to diminished quality of life, disability and even death with a mortality rate 36% depending on the age [1,2]. In clinical practice, the golden standard for the assessment of the disease is dual energy X-ray absorptiometry scan. However, X-ray-based imaging techniques are expensive methods and require radiation exposure. Therefore, the development of non-destructive and non-radiating methods, appropriate to assess the mechanical and structural properties of bone, is crucial.

To this end, several research groups worldwide have investigated the diagnostic and monitoring role of quantitative ultrasound in osteoporotic and fractured bones using experimental, numerical and theoretical means [3,4,5,6,7,8,9]. The research interest in numerical and theoretical methods has allowed the study of phenomena which cannot be observed via experimental measurements extending thus our understanding on the underlying ultrasound propagation mechanisms and the interaction with biological tissues [10]. Specifically, as ultrasound propagates in composite media such as osseous tissues attenuation mechanisms evolve including absorption and scattering phenomena. In soft tissues the viscous forces between neighboring particles moving with different velocities are major sources of acoustic wave absorption [1]. Scattering is induced by the porous nature of cortical bone and the incident energy is partially transferred to the scattered fields. In soft tissues like blood and marrow, ultrasound bulk shear waves are usually ignored as shear waves are highly attenuated at ultrasonic frequencies, while in hard osseous tissues both compression and shear waves are considered.

In addition, the availability of advanced computer-aided computational tools in combination with imaging data derived from micro-computed tomography (μCT) and scanning acoustic microscopy (SAM) have accelerated the evolution of the numerical research [2,3,7,8]. The axial and through transmission techniques have been mainly used to estimate the first arriving signal (FAS) velocity and attenuation as well as to study the propagation of guided waves. In the literature, different thresholds for the detection of the FAS and methodologies to extract guided wave features have been proposed to interpret ultrasound propagation phenomena occurring not only at the periosteal region, but also at deeper bone layers [4,5,6,7,11,12]. More recently, ultrasound reflection and backscattering parameters have been measured to investigate bone microstructure [13,14].

Concerning the first two-dimensional (2D) computational studies, the cortical bone was modeled as a linear elastic homogeneous plate and ultrasound simulations were performed for different plate thicknesses to investigate the thinning of the cortical cortex due to osteoporosis [15,16]. In [17], X-ray computed tomography reconstructions were derived from the human radius in order to develop more realistic 3D computational models of cortical bone. It was shown that for plate thicknesses larger than the wavelength the FAS corresponds to a lateral wave which propagates at the bulk longitudinal velocity of bone, while for very thin plates the FAS wave propagates as the lowest-order symmetric plate mode. More recent 2D and 3D computational studies [2,8,14,17,18,19] take into consideration cortical microstructure, porosity and anisotropy. In [18], SAM images of human femoral neck were used to develop numerical models which account for the sample's overall shape, microstructure, cortical porosity, heterogeneous matrix elasticity and density. It was found that the FAS velocity is not influenced by trabecular bone properties or by the heterogeneities of the cortical bone mineralized matrix. On the other hand, the FAS was sensitive to variations in cortical porosity. In addition, in [2], SAM human femoral neck cross-sections were used to develop multivariate models for the prediction of pore size, porosity, and cortical thickness. It was shown that the FAS velocity decreases with increasing porosity, while an increase of the cortical thickness and pores' diameter leads to an increase in FAS velocity. It was also reported that cortical porosity has the strongest effect on the prediction of the FAS velocity. In addition, a 3D micromechanical model consisting of an anisotropic matrix pervaded by cylindrical pores was developed in [19]. It was found that, for the elderly population, the elastic properties of the mineralized matrix do not undergo large variations among different samples, while changes in the intracortical porosity account for most of the variations of mesoscopic elasticity [19].

In this work, we present a 2D computational study of ultrasonic propagation on healthy and osteoporotic cortical bone to examine the effect of porosity on the FAS velocity. We account for the occurrence of large basic multicellular units, simply called non-refilled resorption lacunae (RL), as the predominance of resorption causes an increase in Haversian canal size leading to the increase of cortical porosity and fragility [14,20]. Initially, various microstructural models are established, mimicking normal and pathological tissue states with porosity of 5%, 10% and 16%. Then, numerical models are presented to investigate the impact of changes in the number, size, position and depth of the RL on the FAS. The central excitation frequencies 0.5 and 1 MHz are examined. To our knowledge, this is the first systematic and parametric numerical study which evaluates the FAS velocity variation along cortical bone derived from small, serial propagation paths for different porosities and frequencies, including the occurrence of pores with different sizes.

2. VALIDATION OF THE NUMERICAL METHOD: A BENCHMARK PROBLEM

The most popular numerical methods for ultrasonic evaluation of cortical bone are the Finite Difference, Finite Element and Boundary Element methods which solve partial differential equations. According to [21], for time domain modeling of broadband or high-frequency waves these methods can become cumbersome and slow due to the requirements for many grid points per wavelength and small time-steps to minimize unwanted numerical dispersion. In order to validate the finite-difference time-domain (FDTD) code that is used in this work (SimSonic, LIP, Paris, France [22,23]), a benchmark problem was established. We aim to compare the outcome of the FDTD method with a MATLAB (Version, The Mathworks, Inc., Natick, MA, USA) toolbox which combines pseudo-spectral and k-space methods, so the spatial gradients are calculated by using Fast Fourier Transforms rather than by using a finite difference stencil (k-Wave, [21]). The computational model ofFigure 1a was used corresponding to an intact cortical plate with thickness of 4 mm and length 40 mm surrounded by water. The material properties of bone and water are presented in Table 1. The grid size was set to 20 μm and the wave propagation simulation time was 10 μs. A point source and receiver were used to calculate the velocity at a distance of 12.5 mm between the transducers. The transducers were placed directly onto the cortical surface. A 1-MHz Gaussian pulse was used as the excitation signal. According to the results of Figure 2, the two waveforms have the same time of arrival of the FAS as well as the same qualititative behavior over time enhancing the accuracy of the FDTD method.

In order to provide quantitative information for the differences of the two waveforms, a convergence study is conducted for the grid sizes of 50, 30 and 20 μm. The methodology of waveform rectification was used followed by numerical integration via the trapezoidal method. A full-wave rectifier converts the examined waveform to a waveform of constant polarity (positive or negative). Specifically, the excitation signal was rectified as well as the received waveforms from k-Wave and Simsonic. Numerical integration was applied using the trapezoidal method to derive the emitted and received signal area. Next, we calculated the ratio of the emitted signal area to the received waveforms' area, which is a quantitative indicator reflecting the whole area of interest. If we denote the calculated ratio with E_{sim} for Simsonic and E_{kw} for k-Wave, the corresponding values were: (a) $E_{kw} = 1166$, $E_{sim} = 1084$ ($\text{difference}_{50\mu m} = 82$) for 50 μm, (b) $E_{kw} = 1935$, $E_{sim} = 1921$ ($\text{difference}_{30\mu m}=14$) for 30 μm and (c) $E_{kw} = 2924$, $E_{sim} = 2919$ ($\text{difference}_{20\mu m} = 5$) for 20 μm.

3. NUMERICAL EVALUATION OF CORTICAL POROSITY USING ULTRASONIC TECHNIQUES

3.1. Model Geometry

In this section, we present the structural features of the computational models representing different cases of cortical porosity and pores' sizes as well as different number and distributions of RL. The structural and material properties were derived from [14], in which the Haversian canals of normal size were differentiated from the RLs to distinguish healthy bones (with no or few RL) from degenerated ones. As can be seen in Figure 1, the cortical bone was modeled as a 2D plate (length 40 mm, width 4 mm) surrounded by water (upper surface 4 mm, lower surface 2 mm).

Two sets of numerical simulations were performed in order to evaluate: (a) porosity changes from 0% to 16% covering the whole region of cortical bone (Series I_Po0-16), (b) the occurrence of a single or a cluster of RL in small propagation paths (Series II_RL). The selection of the porosities of Series I_Po0-16, and the sizes of the RL were based on previous studies investigating the interaction of ultrasound with cortical porosity [2,5,10,14]. However, it should be emphasized that the evaluation of cortical porosity depends on several parameters such as age, gender and region of interest [5]. The first set is presented in Figure 1, in which cortical bone was modeled both as a homogeneous and nonhomogeneous medium to account for the cases of: (a) intact bone (Figure 1a); (b) porosity 5% (Figure 1b); (c) porosity 10% (Figure 1c); and (d) porosity 16% (Figure 1d–f). In Table 2, we present the exact structural features concerning the porosity, number and dimensions of the normal and larger pores. For each microstructural scenario of Table 2, a total of three numerical models were established and the mean local FAS velocity values were calculated as well as the standard error.

Circular scatterers were considered in the transverse direction and the selected radii of normal pores and RL are presented in Table 2. Given the porosity and the dimensions of the plate and scatterers, the number of the pores can be calculated. An algorithm was developed to generate the osseous and soft tissues as greyscale values and the "porosity maps" with a random distribution of the pores. The scatterers were not in contact and did not merge with each other as well as with the cortical boundaries.

Figure 3 illustrates cortical segments of Series II_RL in which the cortex of a long bone was modeled initially as a 2D homogeneous plate including the presence of 1 RL at different lengths, depths and diameters (Figure 3a–e). Then, in Figure 3f–h, we account for cortical microstructure to simulate the case of normal remodeling (Table 2: cases B_Po5_RL1, B_Po5_RL3, B_Po5_RL5) including a single or a cluster of RL in the center of the plate.

3.2. Material Properties

The cortical bone was modeled as an isotropic, linear elastic medium. We assumed that the soft tissues surrounding the cortical plate as well as the circular pores are composed of water. Table 1 summarizes the material properties assigned to the cortical bone, scatterers and soft tissues which were derived from [14].

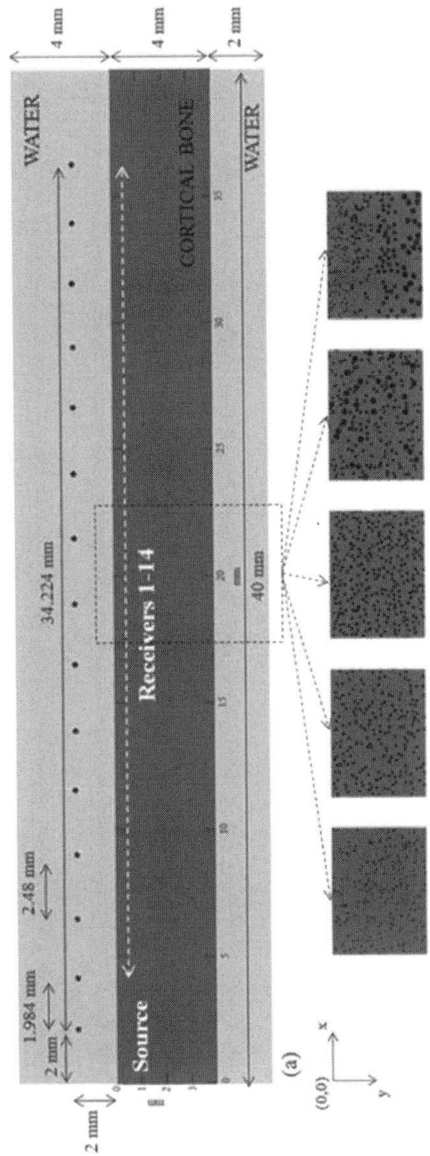

Figure 1. Numerical models of cortical bone corresponding to the first set of simulations (Series I_Po0-16) and Table 2, namely: (**a**) B_Hom; (**b**) B_Po5; (**c**) B_Po10; (**d**) B_Po16; (**e**) B_Po16_RL; and (**f**) B_Po16_Gradual. For each geometry, only one of the three random porosity distributions is depicted in the form of 5 mm cortical segments derived from the original 40mm × 4mm plates. The ultrasound configuration is also presented in (**a**).

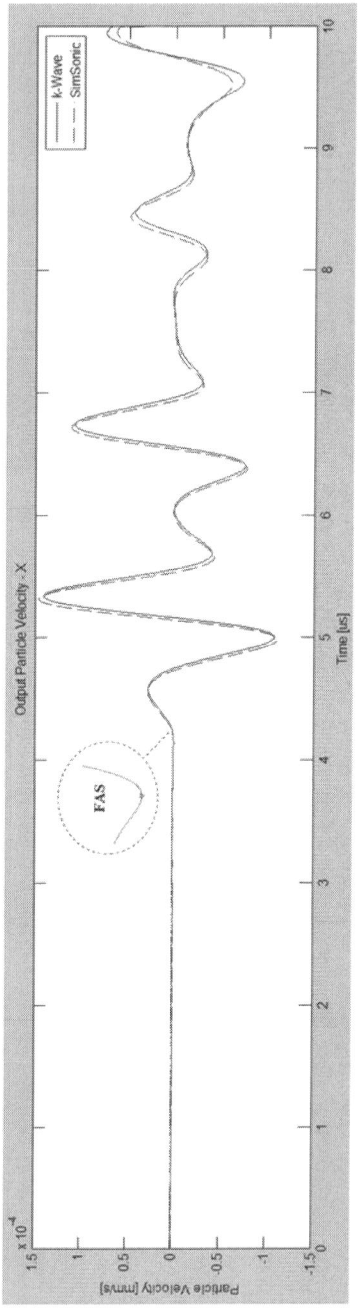

Figure 2. Particle velocity waveforms derived from two different numerical tools, Simsonic and k-Wave, for grid size 20 μm. The threshold for the detection of the FAS is also depicted.

Table 1. Elastic coefficients and density used for the simulations [14].

Propagation Medium	C_{11} (GPa)	C_{12} (GPa)	C_{66} (GPa)	P (g/cm^3)
Water	2.25	2.25	0	1.00
Bone	28.71	10.67	9.02	1.85

Table 2. Examined cases in the tangential direction for the second set of the simulations.

	Description	Porosity (%)	No. of Pores	Pores' Radius (μm)	No. of RL/RLs	Radius of RL/RLs (μm)
B_Hom	Homogeneous bone	0	0	–	–	–
B_Po5	Porous bone, normal pores	5	1592	40	–	–
B_Po5_RL1	Porous bone, normal pores and 1 RL	5	1592	40	1	115
B_Po5_RL3	Porous bone, normal pores and 3 RLs	5	1592	40	3	115
B_Po5_RL5	Porous bone, normal pores and 5 RLs	5	1592	40	5	115
B_Po10	Porous bone, normal pores	10	1592	60	–	–
B_Po16	Porous bone, normal pores	16	2263	60	–	–
B_Po16_RL	Porous bone, normal pores and RLs	16	1400	60	235	115
B_Po16_Gradual	Gradual distribution of the pores, normal pores and RLs	16	1775	40, 60, 80	192	115

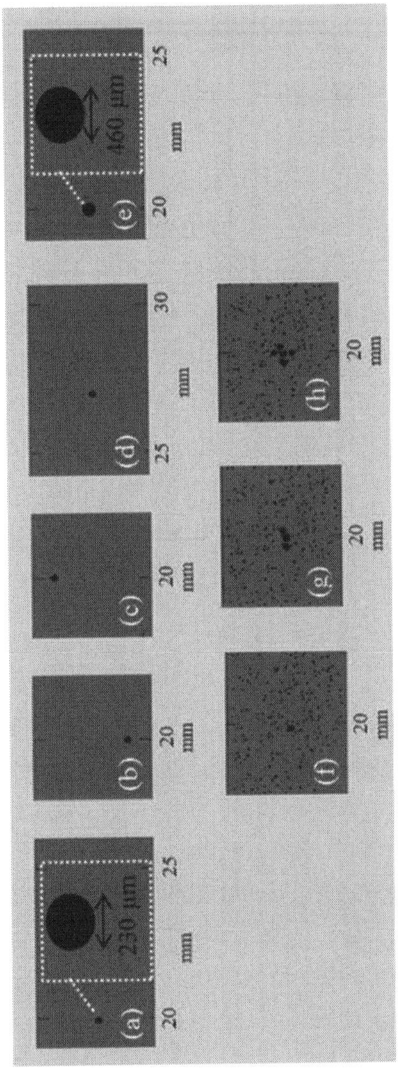

Figure 3. Numerical models of cortical bone corresponding to the second set of simulations (Series II_RL) focusing on the regions of the occurrence of RL: (**a**) homogeneous bone, 1 RL with center coordinates $(x, y) = (20, 2)$; (**b**) homogeneous bone, 1 RL with center coordinates $(x, y) = (20, 3)$; (**c**) homogeneous bone, 1 RL with center coordinates $(x, y) = (20, 1)$; (**d**) homogeneous bone, 1 RL with center coordinates $(x, y) = (27, 2)$ mm; (**e**) homogeneous bone, 1 RL with center coordinates $(x, y) = (20, 2)$ and double the diameter; (**f**) B_Po5_RL1; (**g**) B_Po5_RL3; (**h**) B_Po5_RL5.

3.3. Ultrasound Configuration

Simulations of ultrasound propagation were performed in the tangential direction (Figure 1a). Calculations were conducted by placing one point source and 14 point receivers: (a) at a distance 2 mm from the upper cortical layer (non-contact transducers); (b) directly onto the cortical cortex (implanted transducers). The distance between two successive receivers was 2.48 mm implying that the distance between the emitter and the receivers ranges from 1.984 to 34.224 mm (receiver 1 (R1) – receiver 14 (R14)). The width of each transducer was equal to 11 elements. The use of a linear array of transducers is in agreement with exsisting ultrasonic devices for bone characterization such as the bidirectional device in [24].

3.4. Determination of the Ultrasonic Wave Propagation Path and Velocity

The "local FAS velocities" correspond to small propagation paths and were calculated as the difference in the FAS arrival time of the signals from successive receivers (distance 2.48 mm). In this way, we intend to estimate the structural changes along small regions of cortical bone due to porosity variations or to the presence of larger pores. When multiple receivers are considered, the change in the FAS arrival time Δt (μs) can be calculated using $\Delta t(x) = t(x_i) - t(x_{i-1})$, where i is the number of the receiving position, $t(x_i)$ is the arrival time of the signal at receiver i and $t(x_{i-1})$ is the arrival time at the previous receiver [25]. For the detection of the FAS a threshold was applied to the receiving waveforms corresponding to the identification of the first signal extremum (Figure 2) [26].

3.5. Boundary Conditions

The cortical cortex was surrounded by soft tissues and perfectly matched layers (PML) were used to limit spurious reflections from the boundaries simulating an infinitely long model. The PML efficiency was set to 80 dB which means that the wave reflected by the PML is expected to be 80 dB below the amplitude of a incident wave, in the case of normal incidence [22,23].

3.6. Numerical Simulation and Signal Analysis in the Time Domain

The numerical solution of the 2D wave propagation problem was carried out by using an FDTD code [22,23]. The computations are based on a system of elastodynamic equations corresponding to the propagation of mechanical waves in continuous media which obey Hooke's law and are expressed as:

$$\rho(x)\frac{\partial u_i}{\partial t}(x,t) = \sum_{j=1}^{d}\frac{\partial T_{ij}}{\partial x_j}(x,t) + f_i(x,t) \tag{1}$$

$$\frac{\partial T_{ij}}{\partial t}(x,t) = \sum_{k=1}^{d}\sum_{l=1}^{d} c_{ijkl}(x)\frac{\partial u_k}{\partial x_l}(x,t) + \theta_{ij}(x,t) \tag{2}$$

where subscripts i={1,…,d} refer to the direction of space and d to the space dimension (*d* =2 for 2D computational models), x and t are the space and time variables, $\rho(x)$ is the mass density and c(x) is the fourth-order rigidity tensor. These parameters entirely define the material properties and geometry of the medium. u_i (x, t) are the vector components of the particle velocity field, T_{ij} (x, t) are the components of the stress tensor, while f_i denote the vector components of force sources and θ_{ij} denote the tensor components of strain rate sources.

A Hanning pulse was used as the excitation signal including four sinusoidal cycles [27]. The examined central angular frequencies were 0.5 and 1 MHz. The duration of the simulations was 25 μs. The accuracy of the solution depends on the relation between the element size and the wavelength. According to the software requirements, the stability condition is defined as:

$$\Delta t \leqslant \frac{1}{\sqrt{d}} \frac{\Delta x}{c_{max}} \tag{3}$$

where Δt and Δx denote the time and spatial steps used to approximate time or spatial derivatives, c_{max} is the largest speed of sound amongst all the simulated materials ($c_{max} = c_{bone} = 3939$ m/s), and d refers to the space dimension ($d = 2$). The grid size was set to 8 μm, and a convergence study is presented in the following subsection to demonstrate the accuracy of the findings.

3.6.1 Convergence Study

The grid size must be dense enough to ensure an accurate numerical solution and to minimize the computational cost. In addition, it must be small enough in comparison to the diameter of the pores. The convergence study is based on the numerical model B_Po5 which integrates the smallest scatterers among all the geometries of Table 2. Two transducers were considered which were placed directly onto the cortical surface and their distance was set to 20 mm. FDTD simulations were conducted and the FAS velocities were calculated for the central excitation frequency 1 MHz and the grid step sizes 4, 8, and 16 μm. The calculated FAS values were 3499 m/s for 4 μm, 3495 m/s for 8 μm, and 3489 m/s for 16 μm. It can be concluded that the 8-μm grid size is a good compromise for an accurate numerical solution and less time consuming simulations which is in agreement with [2].

3.7. Statistical Analysis

A statistical analysis was conducted and the results are presented as local mean FAS velocity ± standard error. Linear regression analysis and one-way analysis of variance (one-way ANOVA) were used to evaluate changes in porosity. The statistical findings were considered significant for p-values less than 0.05 [9].

4. RESULTS

Figure 4, Figure 5, Figure 6, Figure 7, Figure 8, Figure 9, Figure 10, Figure 11 and Figure 12 present the results derived from the two simulation sets of Figure 1 and Figure 3 and the central excitation frequencies 0.5 and 1 MHz. As multiple receivers were considered and small propagation paths were examined, the horizontal axis represents the distance of each receiving element from the starting point of the plate with coordinates $(x, y) = (0, 0)$. First, in Subsection 4.1, we present the results derived for porosity changes from 0% to 16%, and then, in Subsection 4.2, the findings corresponding to the occurrence of a single or a cluster of RLs in small propagation paths at different depths, lengths and sizes.

4.1. First Set of Simulations (Series I_Po0-16)

4.1.1. Implanted Transducers

Figure 4 presents the FAS velocity variation profile from multiple implanted receivers if we consider cortical bone as a nonhomogeneous medium with porosity from 0 to 16%. The local

FAS velocities were calculated as the mean values from the same receiving position of three simulation maps with different random distributions of the pores keeping the same microstructure characteristics. The mean FAS velocities and the standard error bars for the excitation frequency 0.5 MHz are presented in Figure 4a–c and for 1 MHz in Figure 4d–f. It can be observed that a porosity increase leads to the decrease of the FAS velocity variation profile, and this is more apparent for the excitation frequency 0.5 MHz.

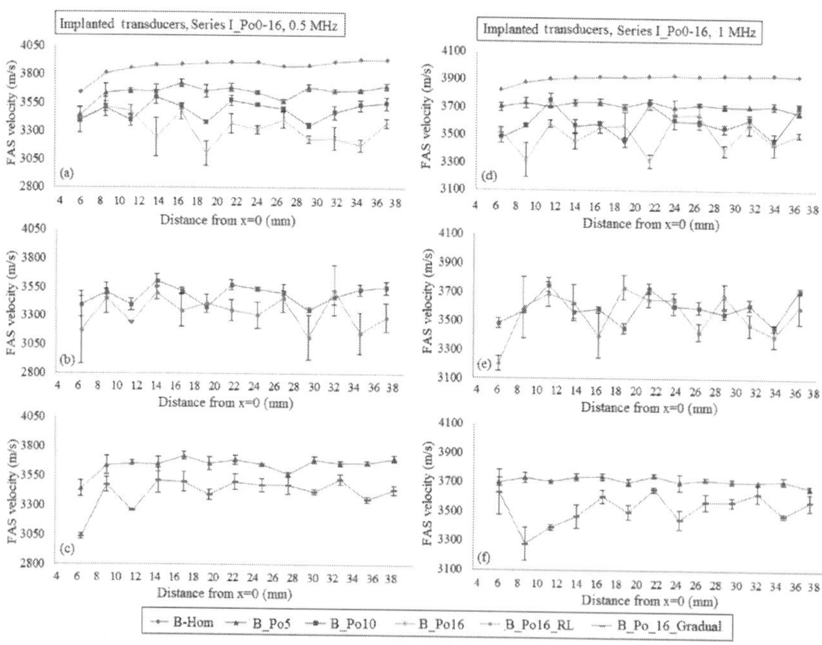

Figure 4. The potential of the mean local FAS velocity to detect changes in cortical porosity from 0%–16%. The diagrams correspond to the first set of simulations under the assumption of implanted transducers and the excitation frequencies: **(a)**–**(c)** 0.5 MHz; and **(d)**–**(f)** 1 MHz. The standard error bars demonstrate the FAS velocity variation for the three numerical models established for each porosity scenario.

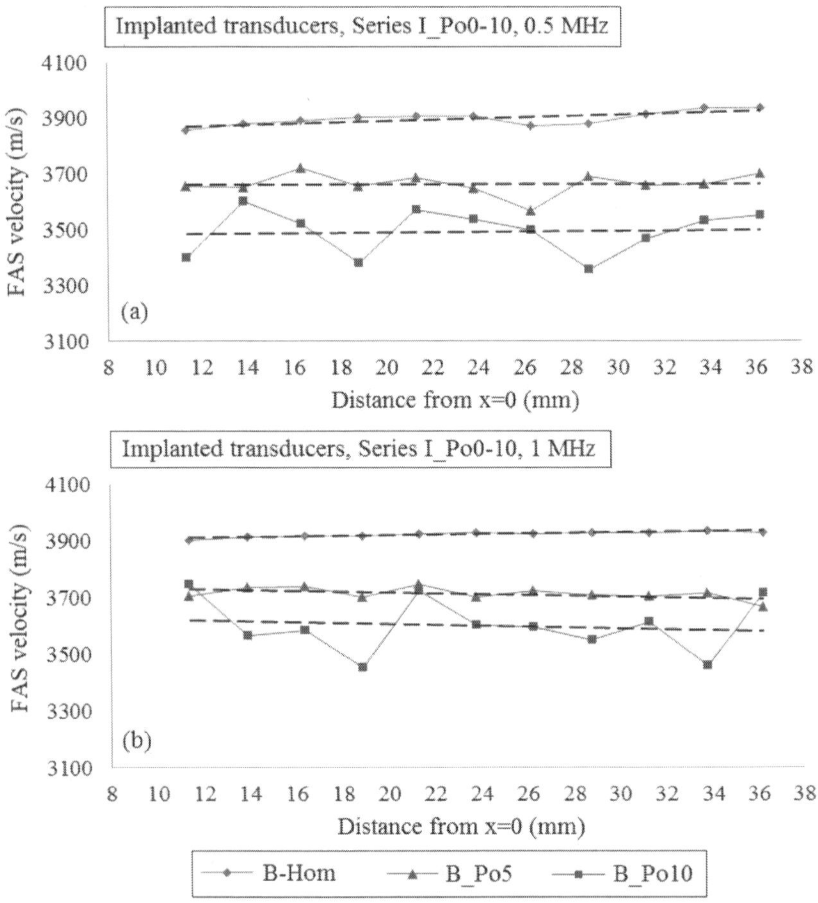

Figure 5. Linear fit derived from multiple receiving positions for cortical porosities from 0 to 10%. The diagrams correspond to the first set of simulations under the assumption of implanted transducers and the excitation frequencies: (**a**) 0.5 MHz; and (**b**) 1 MHz. The corresponding equations, coefficients of determination and root mean square errors are presented in Table 3 and Table 4.

More specifically, if we consider the case B_Hom as the reference case, the relative percentage differences in Figure 4a for the cases B_Po5, B_Po10 and B_Po16 were calculated in the range 4.32%–7.87%, 6.58%–13.53% and 6.41%–20.34%, respectively. These findings can be directly compared with Figure 4d. For the excitation frequency 1 MHz, the relative percentage differences for the cases B_Po5, B_Po10 and B_Po16 were from 3.14% to 6.67%, 3.91%–12.10% and 7.35%–15.56%, respectively. Then, Figure 4b,c,e,f present more realistic scenarios of pathology including both normal pores and RLs for the case B_Po16_RL (Figure 4b for 0.5 MHz and Figure 4e for 1 MHz) and the gradual formation of osteoporosis (Figure 4c for 0.5 MHz and Figure 4f for 1 MHz). In Figure 4b, a porosity increase up to 16% including the occurrence of RLs (B_Po16_RL) leads to a decrease of the local FAS velocities and the relative percentage differences were from 9.23% to 19.71% for the excitation frequency 0.5 MHz. In addition, we observe that there are few positions (e.g. Figure 4b, $x = 30$–32 mm) which show a minor increase of the FAS velocity locally despite the porosity increase. Figure 4e presents the FAS velocities for B_Po16_RL and the excitation frequency 1 MHz. It can be seen that as the frequency increases it is more difficult to discriminate a specific FAS velocity behavior among the cases B_Po10 and B_Po16_RL. Finally, Figure 4c,f show the potential of the local FAS velocity to detect the gradual formation of osteoporosis. The diagram B_Po5 is depicted in the same diagram with B_Po16_Gradual as when osteoporosis is gradually formed the upper cortical layer has the microstructure of a healthy bone, while the porosity and pores' size increase starts from the endosteal region. The relative percentage differences for the case B_Po16_Gradual keeping B_Hom as the reference were calculated from 8.70% to 16.63% for the excitation frequency 0.5 MHz (Figure 4c) and from 5.02% to 15.49% for 1 MHz (Figure 4f), respectively.

Then, one-way ANOVA was conducted and the p-value was calculated. It was found to be lower than 0.05 ($p \approx 0$), implying that there are significant differences between group means. Linear regression analysis was also performed, as in some receiving positions of Figure 4 the mean FAS velocity was found to increase locally despite the porosity increase.Figure 5 shows that linear regression analysis is not influenced by local phenomena, and the examined cases can be discriminated. The corresponding equations, coefficients of determination (R^2) and root mean square errors (RMSE) are included in Table 3 and Table 4. All the results are not presented graphically to reduce complexity, as similar findings toFigure 5 were derived.

4.1.2. Transducers at a Distance of Two Millimeters from the Cortical Cortex

Figure 6 presents snapshots of ultrasound wave propagation for indicative cases considering the excitation frequency 1 MHz and time instant 10 μs. It is shown that, as porosity increases, multiple scattering mechanisms evolve (Figure 6d,e,f). In addition, it can be observed that the application of PML limits spurious reflections from the boundaries, but the phenomenon is not totally diminished.

Then, in Figure 7 the received waveforms are illustrated derived from receiver R8, which is placed above the region of the cluster of 5 RLs (Figure 3h) for indicative cases and the excitation frequency 1 MHz, It can be seen in Figure 7a that the increase of porosity induces a delay in the propagation of the signal if we compare the zero-crossing times. An increase in amplitude was also found with increasing the porosity. A comparison of the waveforms of the cases B_Po16 and B_Po16_RL is presented in Figure 7b, showing that the occurrence of the larger pores has an effect on the received waveforms. However, in Figure 7c, the waveforms for the cases B_Po5 and B_Po5_RL5 almost coincide, while no difference can be observed at the corresponding snapshots of Figure 6.

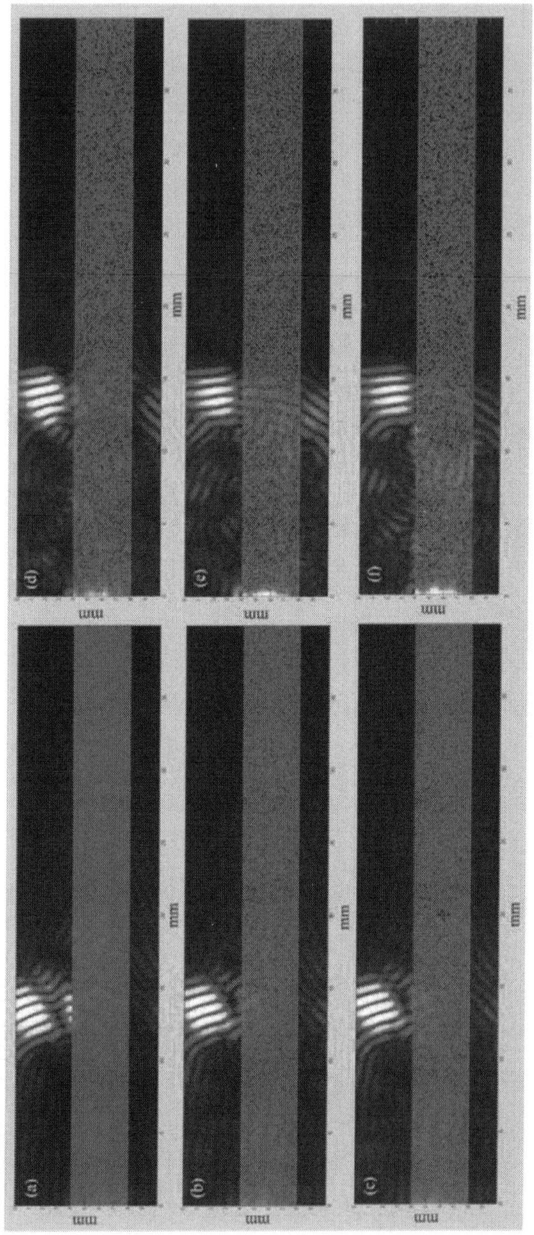

Figure 6. Snapshots of wave propagation for the first set of simulations using non-contact transducers for the excitation frequency of 1 MHz and time instant 10 μs for the cases: (**a**) B_Hom; (**b**) B_Po5; (**c**) B_Po5_RL5; (**d**) B_Po10; (**e**) B_Po16; (**f**) B_Po16_RL.

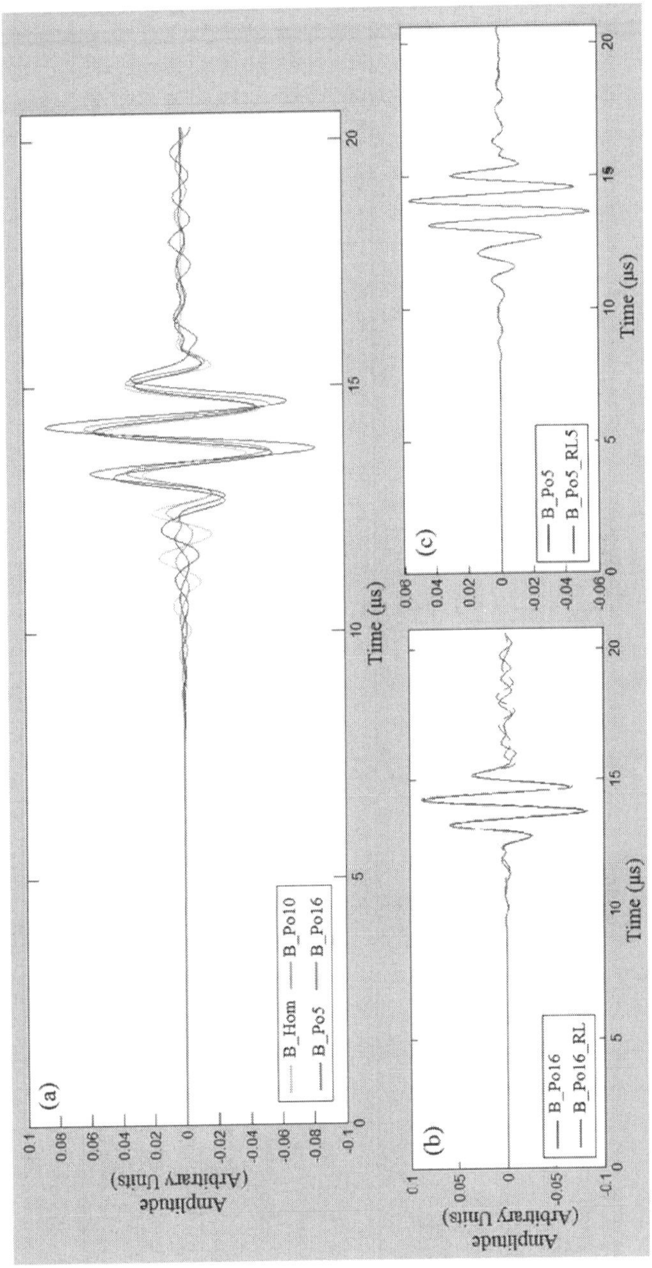

Figure 7. Waveforms derived from receiver R8 for the excitation frequency 1 MHz and simulation time 20 μscomparing the examined cases: (**a**)B_Hom, B_Po5, B_Po10 and B_Po16; (**b**) B_Po16 and B_Po16_RL; (**c**) B_Po5 and B_Po5_RL5.

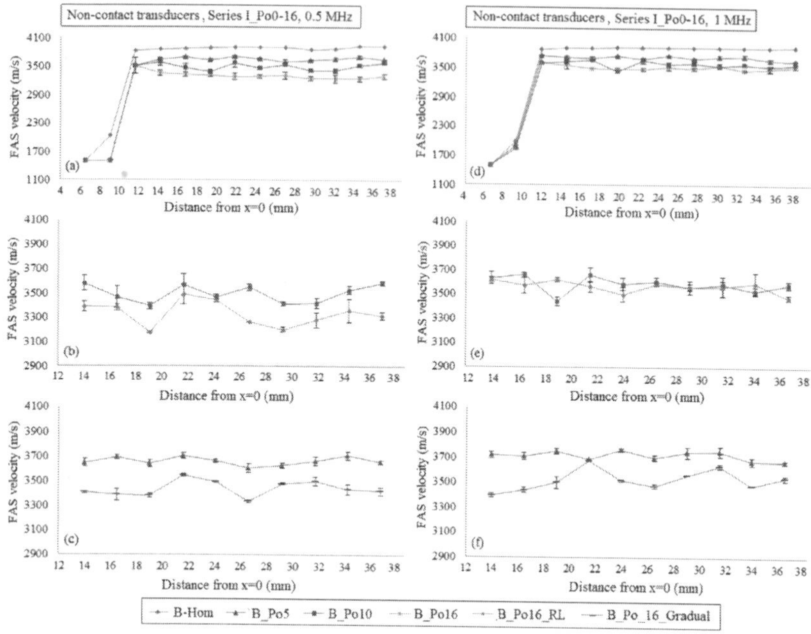

Figure 8. The potential of the mean local FAS velocity to detect changes in cortical porosity from 0% to 16%. The diagrams correspond to the first set of simulations when the transducers are placed at a distance of 2 mm from the cortical cortex. The results for the excitation frequencies : **(a)**–**(c)** 0.5 MHz; and **(d)**–**(f)** 1 MHz are presented. The standard error bars demonstrate the FAS velocity variation for the three numerical models established for each porosity scenario.

Figure 8 presents the mean FAS velocities and standard error bars when the transducers are placed at a distance of 2 mm from the cortical surface. The porosities from 0% to 16% are examined as in Figure 4. In Figure 8a,d, low FAS velocities are observed for the receivers which are placed before x = 12 mm due to the direct propagation of the FAS via the soft tissues.

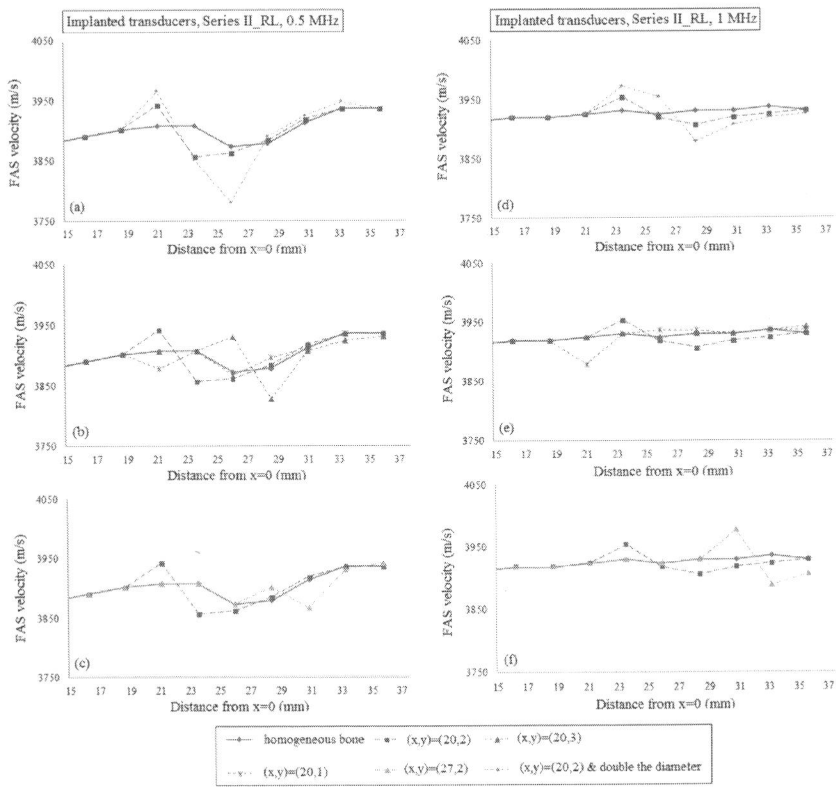

Figure 9. The potential of the local FAS velocity to detect the occurrence of a single RL considering cortical bone as a homogeneous medium (geometries illustrated in Figure 3a–e). The diagrams correspond to the second set of simulations under the assumption of implanted transducers and the excitation frequencies: **(a)**–**(c)** 0.5 MHz; and **(d)**–**(f)** 1 MHz. The legend describes the center coordinates of the RL.

More specifically, if we consider the case B_Hom as the reference case, the relative percentage differences in Figure 8a for 0.5 MHz and B_Po5, B_Po10 and B_Po16 were calculated in the range 4.71%–7.65%, 7.22%–12.83% and 12.90%–17.60%,respectively. For the excitation frequency 1 MHz (Figure 8d), the relative percentage differences for B_Po5, B_Po10 and B_Po16 were from

4.24% to 6.89%, 6.37%–12.19% and 8.78%–11.82%, respectively. Then, Figure 8b,e corresponds to the case B_Po16_RL and are directly compared with the case B_Po10. Keeping B_Hom as the reference case the relative percentage differences for B_Po16_RL were from 10.51 to 18.33% for 0.5 MHz (Figure 8b) and 7.17%–11.62% for 1 MHz (Figure 8e). Thus, even if according to Figure 7b, the two waveforms seem to coincide for propagation time less then 10 μs, however the calculation of the FAS in small propagation paths shows quantitative differences when large pores are considered. The results for the gradual distribution of the pores are shown in Figure 8c,f. The relative percentage differences for B_Po16_Gradual were derived from 9.02% to 14.55% for the excitation frequency 0.5 MHz (Figure 8c) and from 6.31% to 12.95% for 1 MHz (Figure 8f), respectively. It should be mentioned that in Figure 8b,c,e,f, we do not present the whole propagation path, as the initial low values of Figure 8a,b corresponding to the direct FAS propagation through the water were neglected. In addition, linear regression analysis was performed and the results are depicted in Table 3 and Table 4. It was shown that the examined cases can be discriminated even for the diagrams of Figure 8e in which the mean local FAS velocities do not show a specific behavior with increasing the porosity.

Concerning one-way ANOVA, the p-value was lower than 0.05 ($p \approx 0$) ignoring the receiving positions near the source ($d < 12$ mm). The reason for not considering the first receivers in the calculations is that as it can be seen in Figure 8a,d for $d < 12$ mm, the FAS wave propagates directly in the soft tissues and low velocities are derived in comparison to the bulk longitudinal velocity of bone. This assumption was also considered for the cases of Figure 4 to keep the same conditions.

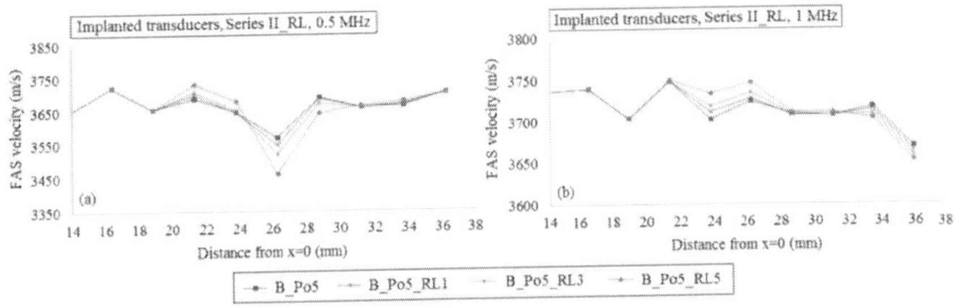

Figure 10. The potential of the local FAS velocity to detect the occurrence of a single or a cluster of RL considering cortical bone as a nonhomogeneous medium for the cases B_Po5, B_Po5_RL1, B_Po5_RL3 and B_Po5_RL5 (geometries illustrated in Figure 3f–h). The diagrams correspond to the second set of simulations under the assumption of implanted transducers and the excitation frequencies: (**a**) 0.5 MHz; and (**b**) 1 MHz.

4.2. Second Set of Simulations (Series II_RL)

4.2.1. Implanted Transducers

This section presents the results corresponding to the occurrence of: (a) a single RL at different depths, lengths and sizes under the assumption of cortical homogeneous bone (Figure 9 and Figure 11); (b) a single or a cluster of RL taking into account cortical microstructure (Figure 10 and Figure 12). In Figure 9 and Figure 10, implanted transducers are used and the excitation frequencies 0.5 and 1 MHz are examined. Figure 9a–c depict the local FAS velocity variation profile for the excitation frequency 0.5 MHz and the presence of a single RL. For comparison purposes, all the diagrams include the calculated values for the intact bone (blue line) and the case of a single RL with center coordinates $(x, y) = (20, 2)$ (red line). All the diagrams start from a distance of 15 mm in the x axis as for lower cortical lengths all the curves coincide. In Figure 9a, we observe that for 0.5 MHz the occurrence of one RL in the center of the plate can be captured by the receivers which are placed directly above the region of interest showing an increase of the FAS velocity of 35 m/s compared to the homogeneous bone diagram. In addition, if we keep the same RL position and double the diameter of the pore (orange line), the receivers which are placed in the position x = 19–21 mm are the first to detect an increase of the FAS velocity of 58 m/s in comparison to the homogeneous bone value. In Figure 9b, we examine how the depth of the RL occurrence influences the FAS velocity. It can be seen that, if we place the RL at a higher cortical surface (grey line), the receivers which first capture the presence of a pore are placed again in the propagation path between x = 19–21 mm showing a velocity decrease of 29 m/s comparing to the reference case of homogeneous bone. On the other hand, if we increase the depth from the cortical surface (purple line), there is a delay in the identification of the RL by the receivers above the region of x = 25–21 mm showing an increase of the FAS velocity of 57 m/s. Finally, Figure 9c shows that if we change the RL position along the x axis, the receivers which are placed above the region of interest are the first to detect an increase of the FAS velocity of 23 m/s revealing the position of the RL formation.

Figure 9d–f presents the local FAS velocity variation profile for the excitation frequency 1 MHz and the presence of a single RL. In Figure 9d, it can be seen that the occurrence of a single RL in the center of the plate (red line) can be identified with a delay by the receivers which follow the region of interest (around x = 23 mm) showing an increase of the FAS velocity of 23 m/s. In addition, it was found that if we double the diameter of the RL (orange line), the same receivers are the first to detect an increase of the FAS velocity of 41 m/s. Then, Figure 9e indicates that by increasing the depth from the cortical surface (purple line), the larger pore cannot be detected as this diagram coincides with the case of homogeneous bone (blue line). On the other hand, the occurrence of the RL near the upper cortical surface (grey line) leads to a decrease of the FAS velocity of 46 m/s derived from the receivers directly above the region of interest. Finally, Figure 9f shows that if we change the position of the RL along the *x* axis, the receivers which are placed approximately at a distance of 3 mm after the pore are the first to capture an increase of the FAS velocity of 47 m/s.

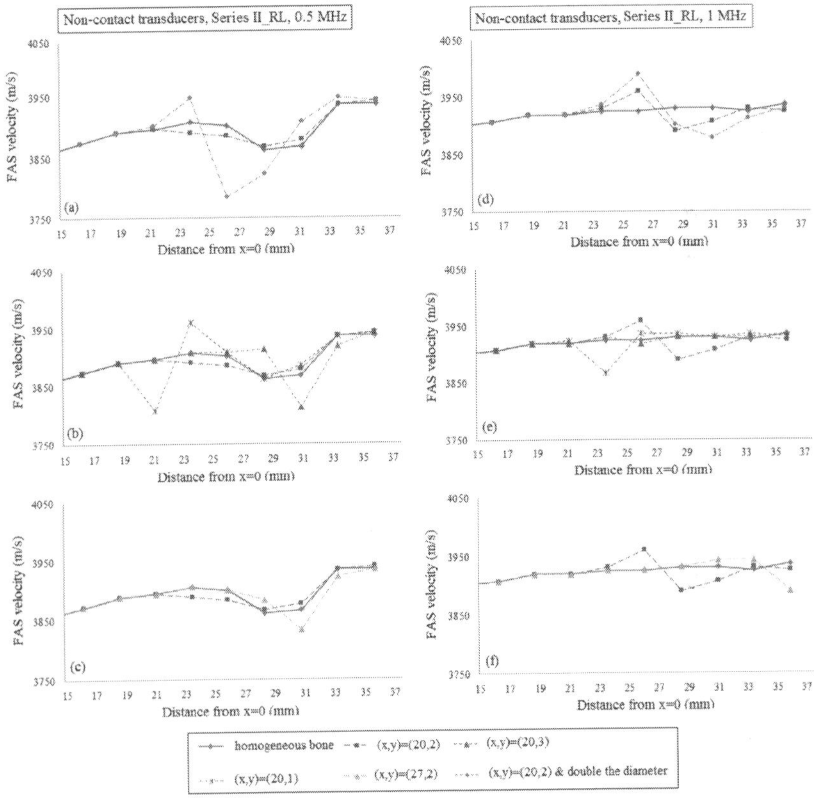

Figure 11. The potential of the local FAS velocity to detect the occurrence of a single RL considering cortical bone as a homogeneous medium (geometries illustrated in Figure 3a–e). The diagrams correspond to the second set of simulations when the transducers are placed at a distance of 2 mm from the cortical cortex. The results for the excitation frequencies : **(a)**–**(c)** 0.5 MHz; and **(d)**–**(f)** 1 MHz are presented. The legend describes the center coordinates of the RL.

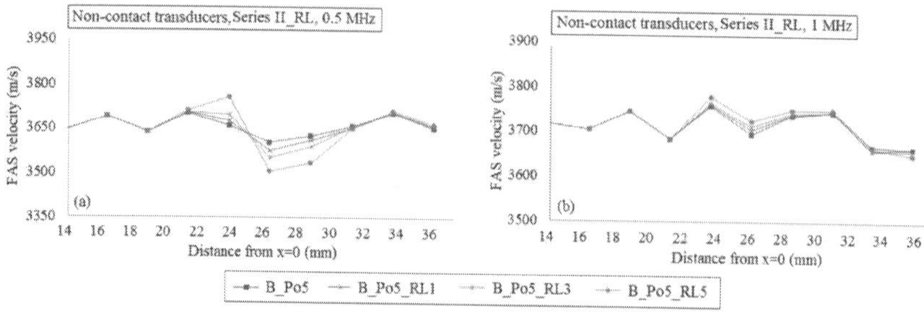

Figure 12. The potential of the local FAS velocity to detect the occurrence of a single or a cluster of RL considering cortical bone as a nonhomogeneous medium for the cases B_Po5, B_Po5_RL1, B_Po5_RL3 and B_Po5_RL5 (geometries illustrated in Figure 3f–h). The diagrams correspond to the second set of simulations when the transducers are placed at a distance of 2 mm from the cortical cortex. The excitation frequencies: (**a**) 0.5 MHz; and (**b**) 1 MHz are examined.

Then, in Figure 10, we investigate the potential of the local FAS velocity to detect the occurrence of a single or a cluster of three or five large pores in the center of the plate taking into account the nonhomogeneous nature of cortical bone. Figure 10a presents the FAS velocity variation profile for the excitation frequency of 0.5 MHz, while Figure 10b the FAS values for 1 MHz considering the diagram of B_Po5 (red line) as the reference case. For the excitation frequency of 0.5 MHz the relative percentage differences were calculated in the range: (a) 0.04%–0.28% for B_Po5_RL1, (b) 0.04%–1.36% for B_Po5_RL3 and (c) 0.04%–3.08% for B_Po5_RL5. It can be seen that the first receivers which can identify the RL occurrence are placed above the region of interest (x = 20–22 mm). A 2-mm delay in the detection of the larger pores is observed in Figure 10b for 1 MHz as they are first identified from the receivers in the region x = 22–24 mm. Lower relative percentage differences were calculated compared to 0.5

MHz from: (a) 0.05%–0.24% for B_Po5_RL1, (b), 0.23%–0.42% for B_Po5_RL3, and (c) 0.05%–0.85% for B_Po5_RL5.

Table 3. Linear fit of the mean first arriving signal (FAS) velocities derived from all the receiving positions and the excitation frequency 0.5 MHz.

	Implanted Transducers			Transducers at a Distance of 2 mm from the Cortical Surface		
	$y = ax + b$	R^2	RMSE	$y = ax + b$	R^2	RMSE
B_Hom	$y = 2.27x + 3844$	0.54	16.56	$y = 3.00x + 3815$	0.52	22.79
B_Po5	$y = 0.17x + 3661$	0.01	37.94	$y = 2.81x + 3577$	0.17	48.84
B_Po10	$y = 0.72x + 3475$	0.01	78.51	$y = 0.28x + 3489$	0.01	66.92
B_Po16	$y = -4.37x + 3417$	0.09	110.08	$y = -3.64x + 3385$	0.53	24.58
B_Po16_RL	$y = -4.09x + 3434$	0.06	123.45	$y = -3.23x + 3410$	0.06	92.93
B_Po16_Gradual	$y = 1.38x + 3409$	0.02	75.88	$y = 1.57x + 3397$	0.03	59.72

Table 4. Linear fit of the mean FAS velocities derived from all the receiving positions and the excitation frequency 1 MHz.

	Implanted Transducers			Transducers at a Distance of 2 mm from the Cortical Surface		
	$y = ax + b$	R^2	RMSE	$y = ax + b$	R^2	RMSE
B_Hom	$y = 1.08x + 3898$	0.80	4.26	$y = 1.65x + 3879$	0.80	6.37
B_Po5	$y = -1.41x + 3748$	0.28	17.82	$y = -1.81x + 3759$	0.19	29.16
B_Po10	$y = -1.54x + 3639$	0.02	93.23	$y = -1.78x + 3628$	0.06	57.31
B_Po16	$y = -1.36x + 3549$	0.01	100.58	$y = -0.06x + 3510$	0.01	30.84
B_Po16_RL	$y = -5.38x + 3702$	0.13	111.10	$y = -3.21x + 3647$	0.28	36.73
B_Po16_Gradual	$y = 3.96x + 3440$	0.15	72.96	$y = 4.49x + 3406$	0.16	73.61

4.2.2. Transducers at a Distance of Two Millimeters from the Cortical Cortex

Figure 11a–c present the local FAS velocity variation profile for the excitation frequency 0.5 MHz and the presence of a single RL by placing the transducers at a distance of 2 mm from the cortical cortex. All the diagrams include the case of intact homogeneous bone (blue line) and the case of a single RL with center coordinates $(x, y) = (20, 2)$ (red line). In Figure 11a, it can be seen that, for 0.5 MHz, the occurrence of one RL in the center of the plate can be captured by the receivers which are placed directly after the region of interest (between x = 21–23 mm) showing a decrease of the FAS velocity of 17 m/s compared to the homogeneous bone diagram. Then, if we keep the same RL position and double the diameter of the RL (orange line) the same receivers are the first to detect an increase of the FAS velocity of 41 m/s in comparison to the homogeneous bone value. Figure 11b shows the impact of the RL depth on the FAS values. We observe that if we place the RL at a higher cortical layer (grey line), the receivers which first capture the presence of a pore are placed directly above the region of interest showing a velocity decrease of 89 m/s comparing to the reference case of homogeneous bone. However, if we increase the depth from the cortical cortex (purple line), the receivers which are placed at a distance approximately 5 mm from the RL are the first to identify its occurrence showing an increase of the FAS velocity of 51 m/s. Finally, Figure 11c shows that if we change the RL position along the x-axis, the receivers which are placed above the region of interest are the first to capture an increase of the FAS velocity of 23 m/s revealing the position of the larger pore.

Additionally, Figure 11d–f present the local FAS velocity variation profile for the excitation frequency 1 MHz and the presence of a single RL. Specifically, Figure 11d shows that the occurrence of a single RL in the center of the plate (red line) can be identified with a delay by the receivers which follow the region of interest (around x = 23–25 mm) showing an increase of the FAS velocity of 35 m/s.

In addition, it can be observed that if we double the diameter of the RL (orange line) the same receivers are the first to identify an increase of the FAS velocity of 65 m/s. In Figure 11e, we observe that if we increase the depth from the cortical surface (purple line) the larger pore cannot be detected as this diagram almost coincides with the case of homogeneous bone. On the contrary, the occurrence of the RL near the upper cortical surface (grey line) leads to a decrease of the FAS velocity of 57 m/s, and it can be identified with a small delay by the receivers placed above the propagation path between x = 21–23 mm. Finally, Figure 11f shows that, if we change the position of the RL along the x-axis, the receivers which are placed between x = 29–31 mm are the first to detect an increase of the FAS velocity of 17 m/s.

Then, in Figure 12, we investigate the potential of the local FAS velocity variation profile to identify the occurrence of a single or a cluster of three or five large pores in the center of the plate taking into account the cortical microstructure. Figure 12a illustrates the local FAS velocities for the excitation frequency 0.5 MHz, while Figure 12b shows the FAS values for 1 MHz considering B_Po5 (red line) as the reference case. For the excitation frequency 0.5 MHz, the relative percentage differences were calculated in the range: (a) 0.05%–0.75% for B_Po5_RL1, (b) 0.05%–1.41% for B_Po5_RL3 and (c) 0.18%–2.71% for B_Po5_RL5. It can be observed that the first receivers which capture the RL are placed between x = 21–23 mm showing a short delay. A 4-mm delay in the detection of the larger pores can be seen in Figure 10b for 1 MHz and lower relative percentage differences were calculated compared to 0.5 MHz from: (a) 0.04–0.32% for B_Po5_RL1, (b) 0.14–0.46% for B_Po5_RL3, (c) 0.05%–0.79% for B_Po5_RL5, respectively.

5. DISCUSSION

This work presents a 2D parametric and systematic computational study aiming to investigate the effect of cortical porosity on ultrasound wave propagation in healthy and osteoporotic long bones. In comparison to previous studies [2,14,26], we established more realistic scenarios, as this is the first time that the distribution of the pores was randomized including normal pores, as well as pores with diameters larger than the Haversian canal size. Moreover, the analysis was not limited to a small segment of cortical bone with the presence of only one large basic multicellular unit as in [14]. The gradual formation of osteoporosis was also considered by simulating more physiological distributions of pores with larger pores in the endosteal cortex and smaller ones in the periosteum. The excitation frequencies 0.5 MHz and 1 MHz were examined to investigate which frequency is more sensitive to detect changes in cortical porosity and the occurrence of RLs. Numerical simulations of wave propagation were performed in the tangential direction and various microstructural models mimicking normal and pathological tissue states were investigated. In comparison to traditional FAS velocity measurements using one source and one receiver, we performed calculations of "local" FAS values in small successive propagation paths along the cortical cortex in order to establish possible guidelines for the interaction of ultrasound with cortical microstructure.

Initially, in Figure 4a,d, and Figure 8a,d, we examined the FAS velocity variation profile along cortical bone for microstructural models with porosities from 0 to 16%. It was made clear that as the porosity increases, the FAS velocities calculated in small propagation paths decrease for both the examined frequencies. However, the relative percentage differences calculated considering B_Hom as the reference case were higher for 0.5 MHz comparing to 1 MHz. Moreover,Figure 7a shows a delay in the time of arrival with increasing the porosity followed by an amplitude increase. This may be explained by the reflection mechanisms, which enhance the received signal for higher porosities. However, despite the significance of the calculation of the FAS velocities from the first signal extremum, the evaluation of the whole waveform would be a possible direction for future research based on guided wave analysis or on the waveform rectification followed by numerical integration via the trapezoidal method (presented in Section 2).

Additionally, the use of implanted transducers led to higher relative percentage differences compared to the use of non-contact transducers, while a delay in the detection of a single or a cluster of RL was observed for non-contact transducers. This can be attributed to the attenuation mechanisms induced by the presence of the soft tissues in the propagation path. The attenuation mechanisms include the evolution of absorption and scattering phenomena. Scattering is induced in bone due to the presence of the pores while soft tissues are major sources of acoustic wave absorption. Despite the accuracy of the FDTD method, which was demonstrated via a convergence study, a limitation of the present work is the simulation weakness of the software to account for attenuation effects resulting from the viscoelasticity [2]. In addition, it should be mentioned that the use of implanted transducers is considered as an invasive procedure for osteoporosis and was examined only for comparison purposes with the non-contact transducers. In clinical practice, their insertion is easy for fracture healing assessment [12,27], but, for the evaluation of osteoporosis, this configuration is not applied during *in vivo*measurements.

Concerning the case of pathology with porosity of 16%, two additional scenarios were investigated to account for the presence of pores with different diameters as well as for the gradual formation of osteoporosis. It was found that the FAS velocity variation profile shows a decrease along the cortical cortex for both B_Po16_RL and B_Po16_Gradual for 0.5 MHz. On the other hand, the excitation frequency of 1 MHz was not convenient for B_Po16_RL (Figure 4e and Figure 8e) as no specific FAS velocity tendency was observed, while, for B_Po16_Gradual, it was sensitive to the gradual porosity increase. A direct comparison of the received signals from B_Po16 and B_Po16_RL was also included in Figure 7b showing a difference in the received signals due to the occurence of large pores.

Additionally, a statistical analysis was conducted based on the estimation of the standard errors, one-way ANOVA and linear regression analysis. It should be mentioned that the receiving positions near the source ($d < 12$ mm) were ignored as in Figure 8a,d, we observe that for short distances between the transducers, low velocities are calculated, revealing that the FAS propagates directly via water. The calculated p-values were lower than 0.05 indicating that there are statistically significant differences between group means as the porosity increases from 0% to 16%. The low p-values can be explained by the high dispersion of the FAS velocity derived from serial measurements for porosity of 16%. Linear regressions analysis revealed a specific FAS velocity trend even for the cases of Figure 4e and Figure 8e. Nevertheless, low R^2 values were derived implying that this analysis could be used only for supplementary observations following traditional local and successive FAS velocity measurements. The establishment of three numerical models for each case is another parameter which requires further discussion. The use of three models is considered as a good compromise to reduce the computational cost and derive accurate conclusions taking into account the calculated standard error values. More specifically, for all the examined cases after three runs, the maximum FAS velocity variation was far less than 5%, thus we decided not to make more runs to keep calculations and simulations simpler.

Then, the case of normal remodeling was examined including the presence of a single or a cluster of RL. Specifically, it was shown that for implanted transducers and 0.5 MHz, as well as for both the hypothetic case of homogeneous and the realistic of nonhomogeneous bone, the presence of one or more RLs can be identified by the receivers which are placed directly above the region of interest revealing the effectiveness of the FAS wave as an indicator. The same observation was made for all the sizes, numbers, depths and lengths of the RL. However, the use of 1 MHz or the transducers' placement into the water led to the detection of larger pores with a short delay when the RLs are placed in the center of the plate (approximately 4 mm) or near the upper cortical cortex (approximately 2 mm). It should be emphasized that the occurrence of a single RL near the lower surface of cortical bone could not be identified when the excitation frequency 1 MHz was applied irrespective of the transducers' configuration. In addition, it was observed that as the receivers are removed from the region of interest the calculated values gradually approximate the values of healthy bone. Nonetheless, the combination of low and high frequencies for the detection of larger pores could be useful, as low frequencies can detect the region of interest along the x axis and higher frequencies can reveal the depth of the pores along the y-axis. For example, in Figure 9b,e, when the RL is placed near the upper cortical layer (grey line), for both frequencies, we can detect the large pore in the same position implying that the pore is near the cortical upper surface. If we place the RL in the center of the plate (red line), the use of 0.5 MHz frequency leads to the detection of the exact x position of the RL, but 1 MHz shows a delay. This reveals that the pore is placed deeper. Finally, Figure 9e shows that for 1 MHz we cannot detect the RL at $y = 3$ (purple line), indicating in combination with 0.5 MHz findings that a large pore exist, but this pore is positioned near the lower surface of the cortical cortex. Thus, even from the limitations of 1 MHz, we may derive specific information concerning the depth of the RL interpreted in combination with 0.5 MHz findings when the occurrence of large basic multicellular units is a factor of major concern.

A possible explanation for the low sensitivity of 1 MHz is that, when the wavelengths are comparable to or smaller ($\lambda_{0.5MHz} = 7.88$ mm, $\lambda_{1MHz} = 3.94$ mm), the thickness of the cortical cortex (4 mm), the FAS wave propagates as a subsurface, lateral wave which travels at the bulk longitudinal velocity of the medium and cannot reflect changes occurring at deeper cortical layers. The calculation of the wavelenths depends on the frequency and the bone's bulk velocity which is 3939 m/s considering the material properties of Table 1. However, bone velocity may vary from 3500 m/s to 4100 m/s depending on the type and anatomic region of bone as well as cortical thickness. According to the literature [15,28], the use of low excitation frequencies (0.2 MHz) has been chosen to yield information on the correlation of the FAS velocity with changes in cortical thickness due to osteoporosis, while it was reported in [28] that the FAS velocities for high frequencies such as 1 MHz were weakly correlated with cortical thicknesses. Therefore, future research should address the question of which frequency range below 0.5 MHz is more convenient for the assessment of osteoporosis considering the occurrence of pores with different sizes.

Concerning the case of homogeneous bone, it was observed that, for 1 MHz, the FAS velocity along the propagation path increases slightly, approximating the bone's bulk velocity (Figure 9d–f), while this is not the case for 0.5 MHz in Figure 9a–c. The observed behavior at 0.5 MHz can be attributed to the fact that the wavelength is almost twice the cortical thickness implying that more complex wave propagation phenomena evolve and the investigation of the propagation of guided waves could convey significant information. On the other hand, for 1 MHz, the FAS propagates as a lateral wave.

Finally, several assumptions were made which require further numerical research and improvements. First of all, although two dimensional geometries were developed, they incorporated various porosity scenarios. Moreover, changes in cortical thickness in cases of pathology in combination with changes in cortical porosity were not considered. Imaging modalities could be also exploited to establish more realistic computational models [29,30]. In addition, the use of point transducers should be mentioned as in experimental procedures the transducers have a finite size. It was reported in [4] that the choice of using point elements instead of finite-sized ones did not affect significantly the recorded ultrasound velocities. According to our previous study using numerical models from SAM images [31], the application of transducers with a finite size larger than the RL diameter could provide significant information concerning the occurrence of a region of RLs. However, further numerical research is needed to investigate the impact of the transducers' size in relation to the scatterers' size on the results. Furthermore, according to [32], the majority of patients with osteoporosis have changes in the mineralization pattern. In particular, reduction of average mineral content was described when compared with agematched controls or when compared with reference values for healthy adults. Quantitative ultrasound was used in [33] for bone characterization and the role of bone mineral density as a major determinant of acoustic properties was highlighted as well as the significance of density-independent relationships with bone microarchitecture. However, according to [34], the main cause of sound attenuation coefficient in bone is porosity, while matrix stiffness has a minor effect. In addition, it was reported in [19] that in aged women the changes in porosity prevail over those of of matrix elasticity to drive the variations of the bone mesoscopic elasticity.

Nevertheless, the results clearly indicated that the excitation frequency 0.5 MHz is more sensitive to detect changes in cortical porosity and the occurrence of RLs. In addition, it was shown that the calculation of the FAS velocity in small propagation paths could potentially provide significant quantitative information for the early diagnosis of osteoporosis and detection of pathologic regions with larger pores with a high risk for fracture.

6. CONCLUSIONS

In this work, we presented a systematic and parametric computational study on ultrasound wave propagation in healthy and osteoporotic bones. The FAS velocity was derived in small propagation paths to extract quantitative criteria for the diagnosis and assessment of osteoporosis taking into consideration not only total changes in cortical porosity, but also the occurrence of larger pores and various microstructural features. The role of the excitation central frequency and the impact of the transducers' positions on the FAS velocity variation profile were examined. For the first time, the gradual formation of osteoporosis was also considered. It was shown that the local FAS velocities derived from a proper combination of low and high frequencies may contribute to the prevention or monitoring of osteoporosis and reveal potential pathologic regions with larger pores at an early stage of the disease. Further numerical research is suggested on challenges for clinical application and more realistic scenarios using scanning acoustic microscopy images.

ACKNOWLEDGMENTS

This work is part funded by the State Scholarships Foundation and the German Academic Exchange Service and by Deutsche Forschungsgemeinschaft (grant Ra1380/9-1).

AUTHOR CONTRIBUTIONS

Vassiliki T. Potsika worked on all the sections of the paper. Also, she was mainly responsible for the writing of the manuscript. Konstantinos N. Grivas and Theodoros Gortsas contributed to the development of the numerical models in Matlab and provided the algorithms for the random distribution of the scatterers. Gianluca Iori mainly contributed toSection 2 concerning the benchmark problem using both k-Wave and Simsonic. Vasilios C. Protopappas and Kay Raum set the parameters for the development of innovative numerical models based on realistic scenarios. They provided basic guidelines concerning: (a) the final porosities which were examined describing different stages of the pathology and (b) the selection of the proper numerical tools. Also, they contributed to the presentation of the results and structure of the paper. Demosthenes Polyzos and Dimitrios I. Fotiadis contributed to the assessment of the final results and the evaluation of: (a) the progress of the students and their individual as well as joint projects (b) the writing of the manuscript.

CONFLICTS OF INTEREST

The authors wish to declare that there is no conflict of interest.

REFERENCES

1. Laugier, P.; Haïat, G. *Bone Quantitative Ultrasound*; Springer: New York, NY, USA, 2010.

2. Rohde, K.; Rohrbach, D.; Gluer, C.C.; Laugier, P.; Grimal, Q.; Raum, K.; Barkmann, R. Influence of Porosity, Pore Size, and Cortical Thickness on the Propagation of Ultrasonic Waves Guided Through the Femoral Neck Cortex: A Simulation Study. *IEEE Trans. Ultras. Ferroel. Freq. Control* **2014**, *61*, 302–313.

3. Grondin, J.; Grimal, Q.; Engelke, K.; Laugier, P. Potential of first arriving signal to assess cortical bone geometry at the hip with QUS: A model based study. *Ultrasound Med. Biol.* **2010**, *36*, 656–666.

4. Moilanen, P.; Talmant, M.; Bousson, V.; Nicholson, P.H.F.; Cheng, S.; Timonen, J.; Laugier, P. Ultrasonically determined thickness of long cortical bones: two-dimensional simulations of *in vitro* experiments. *J. Acoust. Soc. Am.***2007**, *122*, 1818–1826.

5. Malo, M.K.H.; Rohrbach, D.; Isaksson, H.; Töyräs, J.; Jurvelin, J.S.; Tamminena, I.S.; Kröger, H.; Raum, K. Longitudinal elastic properties and porosity of cortical bone tissue vary with age in human proximal femur. *Bone* **2013**, *53*, 451–458.

6. Xu, K.; Liu, D.; Ta, D.; Hu, B.; Wang, W. Quantification of guided mode propagation in fractured long bones.*Ultrasonics* **2014**, *54*, 1210–1218.

7. Papacharalampopoulos, A.; Vavva, M.; Protopappas, V.; Polyzos, D.; Fotiadis, D.I. A numerical study on the propagation of Rayleigh and guided waves in cortical bone according to Mindlin's Form II gradient elastic theory. *J. Acoust. Soc. Am.* **2011**, *130*, 1060–1070.

8. Machado, C.B.; Pereira, W.C.D.A.; Granke, M.; Talmant, M.; Padilla, F.; Laugier, P. Experimental and simulation results on the effect of cortical bone mineralization in ultrasound axial transmission measurements: A model for fracture healing ultrasound monitoring. *Bone* **2011**, *48*, 1202–1209.

9. Rohrbach, D.; Lakshmanan, S.; Peyrin, F.; Langer, M.; Gerisch, A.; Grimal, Q.; Laugier, P.; Raum, K. Spatial distribution of tissue level properties in human a femoral cortical bone. *J. Biomech.* **2012**, *45*, 2264–2270.

10. Potsika, V.T.; Vavva, M.G.; Protopappas, V.C. Computational Modeling of Ultrasound Wave Propagation in Bone. In*Computational Medicine in Data Mining and Modeling*; Springer: New York, NY, USA, 2014; pp. 349–376.

11. Moreau, L.; Minonzio, J.G.; Talmant, M.; Laugier, P. Measuring the wavenumber of guided modes in waveguides with linearly varying thickness. *J. Acoust. Soc. Am.* **2014**, *135*, 2614–2624.

12. Protopappas, V.C.; Fotiadis, D.I.; Malizos, K.N. Guided ultrasound wave propagation in intact and healing long bones. *Ultras. Med. Biol.* **2006**, *32*, 693–708.

13. Malo, M.K.H.; Töyräs, J.; Karjalainena, J.P.; Isakssonc, H.; Riekkinena, O.; Jurvelin, J.S. Ultrasound backscatter measurements of intact human proximal femurs—Relationships of ultrasound parameters with tissue structure and mineral density. *Bone* **2014**, *64*, 240–245.

14. Bourgnon, A.; Sitzer, A.; Chabraborty, A.; Rohde, K.; Varga, P.; Wendlandt, R.; Raum, K. Impact of microscale properties measured by 50-MHz acoustic microscopy on mesoscale elastic and ultimate mechanical cortical bone properties. In Proceedings of IEEE International Ultrasonics Symposium, Chicago, IL, USA, 3–6 September 2014; pp. 636–638.

15. Nicholson, P.; Moilanen, P.; Kärkkäinen, T.; Timonen, J.; Cheng, S. Guided ultrasonic waves in long bones: modelling, experiment and application. *Physiol. Meas.* **2002**, *23*, 755–768.

16. Bossy, E.; Talmant, M.; Laugier, P. Effect of bone cortical thickness on velocity measurements using ultrasonic axial transmission: a 2d simulation study. *J. Acoust. Soc. Am.* **2002**, *112*, 297–307.

17. Moilanen, P.; Talmant, M.; Nicholson, P.H.F.; Cheng, S.; Timonen, J.; Laugier, P. Ultrasonically determined thickness of long cortical bones: Three-dimensional simulations of *in vitro* experiments. *J. Acoust. Soc. Amer.* **2007**, *122*, 2439–2445.

18. Grimal, Q.; Rohrbach, D.; Grondin, J.; Barkmann, R.; Gluer, C.C.; Raum, K.; Laugier, P. Modeling of femoral neck cortical bone for the numerical simulation of ultrasound propagation. *Ultrasound Med. Biol.* **2013**, *40*, 1–12.

19. Granke, M.; Grimal, Q.; Saïed, A.; Nauleau, P.; Peyrin, F.; Laugier, P. Change in porosity is the major determinant of the variation of cortical bone elasticity at the millimeter scale in aged women. *Bone* **2011**, *49*, 1020–1026.

20. Thompson, D.D. Age changes in bone mineralization, cortical thickness, and haversian canal area. *Calcif. Tissue Int.***1980**, *31*, 5–11.

21. Treeby, B.E.; Jaros, J.; Rendell, A.P.; Cox, B.T. Modeling nonlinear ultrasound propagation in heterogeneous media with power law absorption using a k-space pseudospectral method. *J. Acoust. Soc. Am.* **2012**, *131*, 4324–4336.

22. Bossy, E.; Grimal, Q. Numerical methods for ultrasonic bone characterization. In *Bone Quantitative Ultrasound*; Laugier, P., Haiat, G., Eds.; Springer: New York, NY, USA, 2010; pp. 181–228.

23. Bossy, E.; Laugier, P.; Peyrin, F.; Padilla, F. Attenuation in trabecular bone: A comparison between numerical simulation and experimental results in human femur. *J. Acoust. Soc. Am.* **2007**, *122*, 2469–2475.

24. Muller, M; Moilanen, P.; Bossy, E.; Nicholson, P.; Kilappa, V.; Timonen, J.; Talmant, M.; Cheng, S.; Laugier, P. Comparison

of three ultrasonic axial transmission methods for bone assessment. *Ultrasound Med. Biol.* **2005**, *31*, 633–642.

25. Dodd, S.P.; Cunningham, J.L.; Miles, A.W.; Gheduzzi, S.; Humphrey, V.F. Ultrasound transmission loss across transverse and oblique bone fractures: An *in vitro* study. *Ultrasound Med. Biol.* **2008**, *34*, 454–462.

26. Potsika, V.; Spiridon, I.; Protopappas, V.; *et al.* Computational study of the influence of callus porosity on ultrasound propagation in healing bones. In Proceedings of 36th Annual International Conference of the IEEE Engineering in Medicine and Biology Society, Chicago, IL, USA, 3–6 September 2014; pp. 636–638.

27. Potsika, V.T.; Grivas, K.N.; Protopappas, V.C.; Vavva, M.G.; Raum, K.; Rohrbach, D.; Polyzos, D.; Fotiadis, D.I. Application of an effective medium theory for modeling ultrasound wave propagation in healing long bones. *Ultrasonics* **2014**, *54*, 1219–1230.

28. Moilanen, P.; Nicholson, P.H.F.; Kilappa, V.; Cheng, S.; Timonen, J. Assessment of the cortical bone thickness using ultrasonic guided waves: modelling and *in vitro* study. *Ultrasound Med. Biol.* **2007**, *33*, 254–262.

29. Raum, K.; Grimal, Q.; Varga, P.; Barkmann, R.; Glüer, C.C.; Laugier, P. Ultrasound to Assess Bone Quality. *Curr. Osteoporos. Rep.* **2014**, *12*, 154–162.

30. Raum, K.; Leguerney, I.; Chandelier, F.; Talmant, M.; Saïed, A.; Peyrin, F.; Laugier, P. Site-matched assessment of structural and tissue properties of cortical bone using scanning acoustic microscopy and synchrotron radiation μCT. *Phys. Med. Biol.* **2006**, *51*, 733–746.

31. Potsika, V.T.; Gortsas, T.; Grivas, K.N.; Protopappas, V.C.; Vavva, M.G.; Raum, K.D.; Polyzos, D.; Fotiadis, D.I. Computational modeling of guided ultrasound wave propagation in healthy and osteoporotic bones. In ELEMBIO 6th National Conference, Patras, Greece, 2014.

32. Roschger, P.; Misof, B.; Paschalis, E.; Fratzl, P.; Klaushofer, K. Changes in the Degree of Mineralization with Osteoporosis and its Treatment. *Curr. Osteoporos. Rep.* **2014**, *12*, 338–350.

33. Cortet, B.; Boutry, N.; Dubois, P.; Legroux-Gérot, I.; Cotton, A.; Marchandise, X. Does quantitative ultrasound of bone reflect more bone mineral density than bone microarchitecture. *Calcif. Tissue Int.* **2004**, *74*, 60–67.

34. Sasso, M; Haïat, G.; Yamato, Y.; Naili, S.; Matsukawa, M. Frequency Dependence of Ultrasonic Attenuation in Bovine Cortical Bone: An In Vitro Study. *Ultrasound Med. Biol.* 2007, *33*, 1933–1942.

CHAPTER 3

Unidirectional Wave Propagation in Low-Symmetric Colloidal Photonic-Crystal Heterostructures

Vassilios Yannopapas

Department of Physics, National Technical University of Athens, GR-15780 Athens, Greece; Tel.: +30-210-7721481

ABSTRACT

We show theoretically that photonic crystals consisting of colloidal spheres exhibit unidirectional wave propagation and one-way frequency band gaps without breaking time-reversal symmetry via, e.g., the application of an external magnetic field or the use of nonlinear materials. Namely, photonic crystals with low symmetry such as the monoclinic crystal type considered here as well as with unit cells formed by the heterostructure of different photonic crystals show significant unidirectional electromagnetic response. In particular, we show that the use of scatterers with low refractive-index contrast favors the formation of unidirectional frequency gaps which is the optimal route for achieving unidirectional wave propagation.

KEYWORDS

photonic crystals; wave transmission; multiple-scattering

1. INTRODUCTION

Photonic computing holds promise for superior processing of information in terms of speed, liability and bandwidth. Key ingredients in a photonic/optical computer are the waveguiding devices which steer light among different photonic components (diodes, optical transistor, optical buffers, *etc.*). Although this is easily accomplished in the microwave regime via conventional metallic waveguides, in the optical regime it still is a subject of intensive research as commercially available optical fibers cannot efficiently bend light around sharp corners at micrometer or even lower dimensions. Perhaps the most robust method for steering and bending light in the micrometer scale are the so-called photonic-crystal fibers [1]. Namely, light is localized efficiently within a dielectric or air core due to the suppression of propagation within the photonic-crystal cladding which exhibits an omnidirectional band gap for the waveguiding frequencies. Device performance in photonic-crystal fibers can be deteriorated by fabrication imperfections which lead to important backscattering of light, especially at sharp edges. A different approach for guiding light in photonic computers lies with the discipline of plasmonics wherein the excitation of surface plasmons in noble-metal components leads to light localization in the nanoscale enabling efficient light steering at very small volumes [2]. The main obstacle in this technology are the inherent losses of noble metals in the optical regime which degrade light guiding at sufficiently long distances.

A recent idea which has been put forward as an alternative to the current technologies for light guiding is the so-called unidirectional or one-way waveguiding, *i.e.*, propagation of light solely along one of the two possible directions in a straight waveguide. A direct consequence of the inhibition of light propagation in the other direction is the complete absence of light backscattering due to fabrication-induced disorder and imperfections or at very sharp guiding edges. This is easily achieved in the GHz regime via introducing time-reversal symmetry breaking in the microwave circuitry with the use of ferromagnetic materials and application of magnetic fields [3,4,5,6,7]. However, in the optical regime, again, it is really challenging to implement such a strategy as the magnetic contribution of ferromagnetic materials at this spectral region is very weak. Artificial structures made of magneto-optically responsive materials can potentially provide a means to achieve unidirectional waveguiding without however success up to now, due to their weak contribution to the overall electromagnetic response of such materials. Although the magneto-optical contribution can be very much enhanced with the insertion of plasmonic materials [8,9,10,11,12,13,14,15,16,17] , experimental evidence of unidirectional waveguiding via magneto-optical structures is still lacking.

On the other hand, there have been alternative approaches so as to achieve unidirectional propagation/guiding without the use of magnetic materials and subsequent application of an external magnetic field. Namely, unidirectional propagation has been reported in nonlinear photonic crystals (PCs) [18,19,20,21], in photonic-crystal heterostructures [22,23,24,25] as well as in 2D PCs with periodically modulated interfaces [26,27,28]. However, in the above cases, although the application of an external magnetic field is no longer needed, the range of frequencies and angles of incidence within which unidirectional propagation occurs is very narrow. Recently we have presented a linear, passive design of a purely dielectric 3D PC which possess unidirectional (one-way) photonic band gaps spanning over wide regions in both the Brillouin zone and EM spectrum [29]. The respective unit cell of the PC possess a degree of chirality while, at the same time, consists of different material scatterers giving rise to unidirectional wave propagation and frequency gaps. In the present work, we proceed further with this PC design and explore its potential for unidirectional light propagation. Namely, we consider PC designs with real materials simulating the response of colloidal photonic-crystal heterostructures. We show, in particular, that the emergence of unidirectional frequency bands and gaps is favored by materials with low refractive-index contrasts wherein the PC frequency bands are described as a perturbation to the free-photon frequency bands of an empty lattice.

2. DESCRIPTION OF THE PHOTONIC-CRYSTAL HETEROSTRUCTURE

We deal with the 3D photonic crystal shown in Figure 1. It can be described as a monoclinic crystal with a four-point basis. The crystal is viewed as a succession of planes of spheres parallel to the *xy*-plane. Each plane possesses the same 2D periodicity defined by the primitive vectors $\mathbf{a}_1 = (a, 0, 0)$ and $\mathbf{a}_2 = (0, a, 0)$ [(001) crystallographic surface]. This means that we have a square lattice with lattice vectors $\mathbf{R}_n = n_1\mathbf{a}_1 + n_2\mathbf{a}_2$, $n_1, n_2 = 0, \pm1, \pm2, \cdots$ and corresponding reciprocal lattice vectors $\mathbf{g} = m_1\mathbf{b}_1 + m_2\mathbf{b}_2$, $m_1, m_2 = 0, \pm1, \pm2, \cdots$ where $\mathbf{b}_i \cdot \mathbf{a}_j = 2\pi\delta_{ij}$, $i, j = 1, 2$. A unit layer of the crystal consists of four non-primitive planes of spheres at (0, 0, 0), (0, 0, $a/2$), (0, 0, a), and ($-0.3a$, 0, $3a/2$). All the spheres of the crystal of Figure 1 have the same radius $S = 0.2a$. The spheres of the first layer are of type A, of the second type B while the spheres at the third and fourth layers are of type C. The $(n + 1)$-th unit layer is obtained from the n-th layer by the primitive translation $\mathbf{a}_3 = (-0.3a, 0, 2a)$. Such a monoclinic PC lattice has only one mirror plane [xz-plane, *i.e.*, (010) crystallographic surface] and is therefore a lattice of low symmetry. We can thus expect that light transmission through a finite slab of such a monoclinic PC lattice would be different for incidence from the two opposite faces of the slab, if the slab is considered as a stack of planes parallel to a crystallographic direction other than the (010). In what follows, the different sphere types correspond to spheres made of different materials. In principle, one-way propagation can also be achieved by having spheres of different size or scatterers of dissimilar shape.

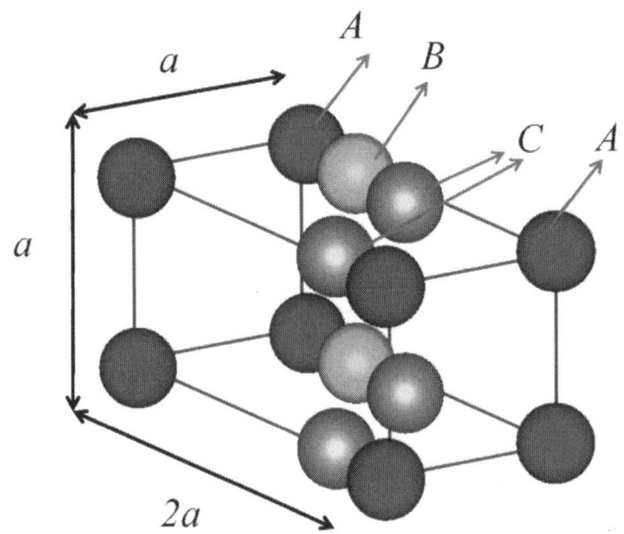

Figure 1. Unit cell of 3D PC with one-way photonic band gaps: monoclinic crystal consisting of four non-primitive planes of spheres parallel to the (001) surface at positions (0, 0, 0), (0, 0, $a/2$), (0, 0, a), and ($-0.3a$, 0, $3a/2$).

3. CALCULATION METHOD

The PC considered in the preceding section will be studied in terms of its frequency band structure and light transmission/absorption by employing the layer-multiple-scattering (LMS) method [30,31,32,33,34]. The LMS method is ideally suited for the calculation of the transmission, reflection and absorption coefficients of an EM wave incident on a composite slab consisting of a number of layers which can be either planes of non-overlapping spherical [30,31,32] or nonspherical axisymmetric [33] and of general shape [35] particles or clusters of such [34], with the same 2D periodicity or homogeneous plates. For each plane of particles, the method calculates the full multipole expansion of the total multiply scattered wave field and deduces the corresponding transmission and reflection matrices in the plane-wave basis. The transmission and reflection matrices of the composite slab are evaluated from those of the constituent layers. By imposing periodic boundary conditions one can also obtain the (complex) frequency band structure of an infinite periodic crystal. The method applies equally well to non-absorbing systems and to absorbing ones. Its main advantage over the other existing numerical methods lies in its efficient and reliable treatment of systems containing strongly dispersive materials such as Drude-like and polaritonic materials.

4. RESULTS AND DISCUSSION

In order to assess the asymmetrical transmission of light which would possibly lead to the generation of unidirectional photonic frequency bands/gaps, we calculate the transmittance of light incident on finite slabs of the 3D PC of Figure 1, based on the LMS method analyzed above. In Figure 2 we show a finite slab of the photonic crystals studied here wherein the slab consists of eight unit layers. As type-A spheres we have considered silica SiO_2 (ϵ = 2.1) spheres, as type-B polystyrene (ϵ = 2.6) spheres and as type-C silicon (ϵ = 11.9) (Si) spheres. All three types of materials have been the basis for the fabrication of colloidal photonic crystals of nanospheres (for a review see [36]).

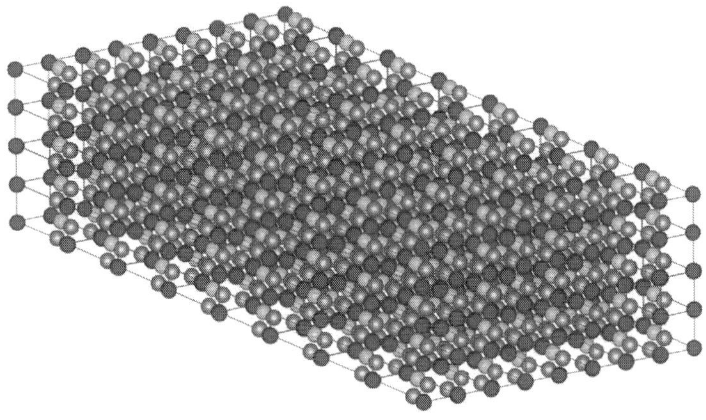

Figure 2. Finite slab of the photonic crystal of Figure 2 consisting of eight unit layers.

T^{+} refers to the transmittance of light incident from the left (001) face of the PC slab, *i.e.*, a wave propagating from left to right whereas T^{-} refers to the transmittance of light incident from the right (001) face of the PC slab, *i.e.*, a wave propagating from right to left. Figure 3 shows T^{+} and T^{-} for light incident off-normally on a finite slab of the Figure 2 with wavevector$\mathbf{k}_{\|} = (0.25, 0)\pi/a$, for s- (a) and p- (c) polarization. The PC slab consists of eight unit layers where each layer contains four square lattices (planes) arranged as shown in Figure 2. Evidently, there are spectral regions where T^{+} and T^{-} differ significantly, such as, e.g., from $\omega a/c = 4.8 - 5.0$ and from $\omega a/c = 5.2 - 5.8$ for *s*-polarized light. For *p*-polarized light, T^{+} and T^{-}differ substantially from $\omega a/c= 5.9 - 6.1$. Despite the important differences in light transmission for left and right incidence a clear case where there is a frequency gap for only one case of incidence direction (left or right) cannot be observed. By inspecting the corresponding frequency band structure (Figure 3b) we see that there exists a single band gap (from $\omega a/c = 5.0 - 5.2$) for both polarizations and directions of incidence (left and right) as well as other gaps for only *s*- or *p*-polarized incident light. We also observe a definite asymmetry in the frequency bands relative to the

$k_z = 0$ which stems from the lack of reflection symmetry at this crystal direction.

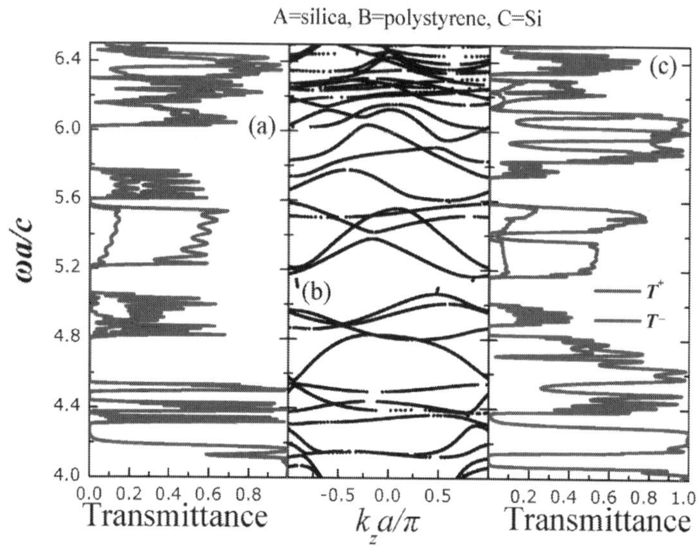

Figure 3. Transmittance for s- (a) and p- (c) polarized light incident with $\mathbf{k}_{\|} = (0.25, 0)\pi/a$ on a finite slab of the PC (see Figure 2) consisting of eight unit layers whereas type-A spheres we have considered silica SiO_2 ($\epsilon = 2.1$) spheres, as type-B polystyrene ($\epsilon = 2.6$) spheres and as type-C silicon ($\epsilon = 11.9$) (Si) spheres. T^+ (T^-) is the transmittance for light incident from the left (right) (001) faces of the slab. (b) Frequency band structure of the infinitely periodic PC of Figure 1 for $\mathbf{k}_{\|} = (0.25, 0)\pi/a$.

Next, we consider a PC heterostructure with the (same) unit cell of Figure 1 but with the polystyrene spheres (the B-type spheres in Figure 3) substituted with germanium ($\epsilon = 16.2$) ones. Evidently, both spectra of T^+ and T^- practically coincide apart from two very narrow regions denoted by the arrows in Figure 4c. This is in accordance with the corresponding frequency band structure of Figure 4b where all frequency bands are symmetric with respect to the center $k_z = 0$. Obviously, the inclusion of materials of high-index contrast in the considered photonic-crystal heterostructures does not favor the emergence of unidirectional propagation.

In an opposite trend (in relation with Figure 4), we substitute the high-index silicon spheres (type-C spheres) of Figure 4with sapphire ($\epsilon = 3.13$) ones [37]—see Figure 5. We clearly see an entirely different picture from that of Figure 4. The frequency bands show a clear asymmetry with respect to $k_z = 0$ which is reflected in the corresponding transmission spectra of Figure 5a,c. Namely, we identify two regions of clear unidirectional frequency gaps. Namely, for both *s*- and *p*-polarized light, there exists a T^+-frequency gap around $\omega a/c = 5.0$ and a T^--frequency gap around $\omega a/c = 5.4$. Asymmetric transmission (for left- and right-incidence) is also evidenced $\omega a/c = 5.8$. We note that the emergence of the unidirectional frequency gaps is not generated by complete absence of frequency bands for one of the k_z-directions ($k_z > 0$ or $k_z < 0$). There is a more complex mechanism which explains the emergence of unidirectional frequency gaps in terms of the matching of the group velocities of incident waves with that of the frequency bands (described analytically in [29]).

The comparison between Figure 4 and Figure 5 leads to the conclusion that the inclusion of high-index spheres does not favor the emergence of unidirectional gaps and propagation. This is a direct consequence of the localized nature of the EM modes of a PC consisting of high-index scatterers. Namely, due to the high refractive index (dielectric constant) of the scatterers the EM field is much localized within a plane of scatterers and wave propagation is achieved by hopping between neighboring planes in a manner resembling the tight-binding picture of electrons in ordinary atomic solids. In accordance with this picture, the lateral (normal to the growth direction of the crystal) displacement of a plane of spheres in an otherwise rectangular arrangement of planes of spheres, see, e.g., the fourth plane of C-type spheres of Figure 1, does not seem to affect seriously the frequency band structure

(Figure 4b) and subsequently the transmission spectrum. On the other hand, if the PC comprises of low-index spheres, the photons propagate nearly freely within the PC (a completely opposite picture than the tight-binding mechanism describing wave propagation in a PC of high-index scatterers) except from undergoing diffraction for the k-wavevectors implied by the underlying lattice symmetry. Since a monoclinic crystal structure yields an entirely different diffraction pattern than a rectangular one, we expect that the lateral displacement of the fourth plane (of C-type spheres) of Figure 1 to have a large impact on wave propagation of nearly free photons.

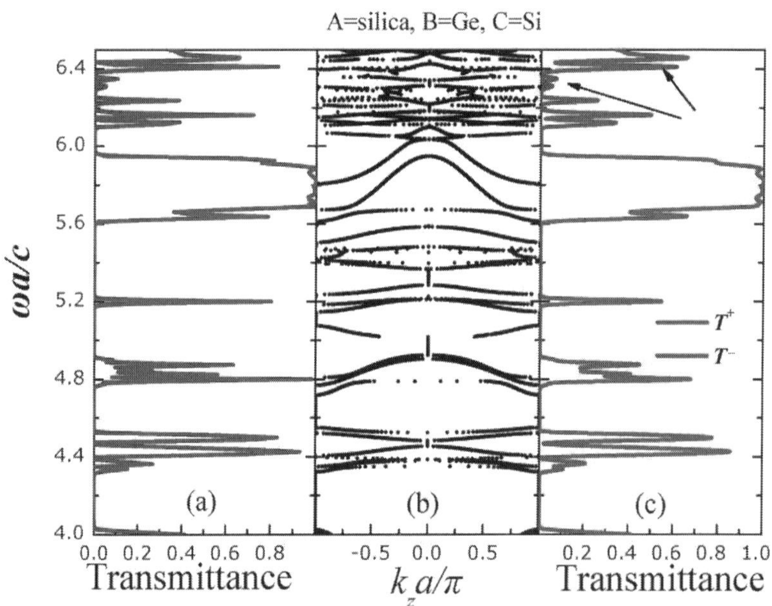

Figure 4. Transmittance for s- (**a**) and p- (**c**) polarized light incident with $\mathbf{k}_{\|} = (0.25, 0)\pi/a$ on a finite slab of the photonic crystal (PC) (see Figure 2) consisting of eight unit layers whereas type-A spheres we have considered silica SiO_2 ($\epsilon = 2.1$) spheres, as type-B germanium ($\epsilon = 16.2$) spheres and as type-C silicon ($\epsilon = 11.9$) (Si) spheres. T^+ (T^-) is the transmittance for light incident from the left (right) (001) faces of the slab; (**b**) Frequency band structure of the infinitely periodic PC of Figure 1 for $\mathbf{k}_{\|} = (0.25, 0)\pi/a$.

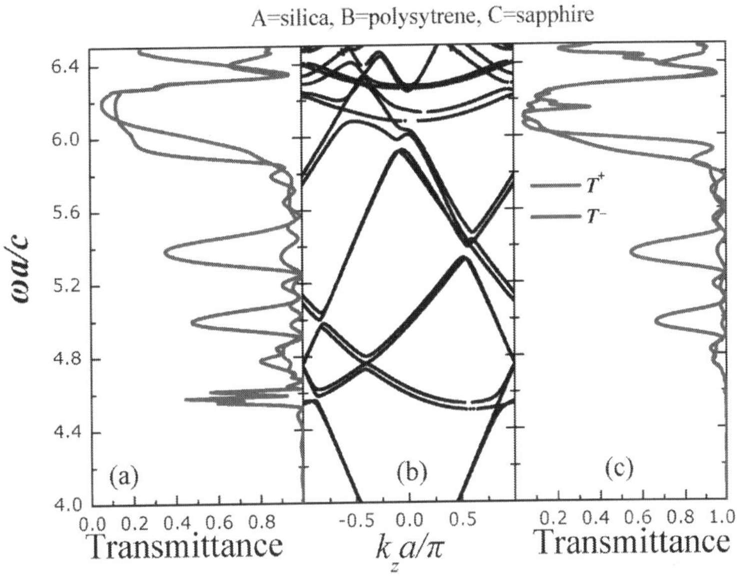

Figure 5. Transmittance for s- (**a**) and p- (**c**) polarized light incident with $\mathbf{k}_{\|} = (0.25, 0)\pi/a$ on a finite slab of the PC of Figure 1 consisting of eight unit layers whereas type-A spheres we have considered silica SiO_2 ($\epsilon = 2.1$) spheres, as type-B polystyrene ($\epsilon = 2.6$) spheres and as type-C sapphire ($\epsilon = 3.13$) (Si) spheres. T^{+} (T^{-}) is the transmittance for light incident from the left (right) (001) faces of the slab; (**b**) Frequency band structure of the infinitely periodic PC of Figure 1 for $\mathbf{k}_{\|} = (0.25, 0)\pi/a$.

5. CONCLUSIONS

We have shown numerically that it is possible to obtain unidirectional EM wave propagation with plain, linear, dielectric photonic crystals consisting of colloidal micro- or nanospheres without applying external magnetic fields or using nonlinear materials. Prerequisites for observing unidirectional wave propagation are the low symmetry of the photonic crystals and the use of a heterostructure of photonic crystals of different scatterers (being different in material composition and/or size) as the unit cell. We have explored photonic crystals of different material composition wherein we have found that efficient unidirectional propagation in the form of unidirectional frequency gaps/bands emerges when scatterers of low-index contrast are used within the heterostructured unit cell of the crystal. For a practical realization, say, in the optical regime, of a low-symmetric photonic crystal such as those considered here, structures with lattice constant of about 480 nm are needed for a unidirectional band gap around 600 nm, which are usually fabricated via colloidal self-assembly [38,39,40,41].

CONFLICTS OF INTEREST

The authors declare no conflict of interest.

REFERENCES

1. Zolla, F.; Renversez, G.; Nicolet, A.; Kuhlmey, B.; Guenneau, S.; Felbacq, D. *Foundations of Photonic Crystal Fibres*; Imperial College Press: London, UK, 2005.

2. Bozhevolnyi, S. *Plasmonic Nanoguides and Circuits*; Pan Stanford Publishing: Singapore, 2009.

3. Haldane, F.D.M.; Raghu, S. Possible realization of directional optical waveguides in photonic crystals with broken time-reversal symmetry. *Phys. Rev. Lett.* **2008**, *100*, 013904.

4. Wang, Z.; Chong, Y.D.; Joannopoulos, J.D.; Soljacˇic´, M. Reflection-free one-way edge modes in a gyromagnetic photonic crystal. *Phys. Rev. Lett.* **2008**, *100*, 013905.

5. Wang, Z.; Chong, Y.D.; Joannopoulos, J.D.; Soljacˇic´, M. Observation of unidirectional backscattering-immune topological electromagnetic states. *Nature* **2009**, *461*, 772–775.

6. Yu, Z.; Veronis, G.; Wang, Z.; Fan, S. One-way electromagnetic waveguide formed at the interface between a plasmonic metal under a static magnetic field and a photonic crystal. *Phys. Rev. Lett.* **2008**, *100*, 023902.

7. Ao, X.; Lin, Z.; Chan, C.T. Dirac spectra and edge states in honeycomb plasmonic lattices. *Phys. Rev. B* **2009**, *80*, 033105.

8. Li, Y.; Zhang, Q.; Nurmikko, A.V.; Sun, S. Enhanced magnetooptical response in dumbbell-like Ag – $CoFe_2O_4$nanoparticle pairs. *Nano Lett.* **2005**, *5*, 1689–1692.

9. González-Díaz, J.B.; García-Martín, A.; García-Martín, J.M.; Cebollada, A.; Armelles, G.; Sepúlveda, B.; Alaverdyan, Y.; Käll, M. Plasmonic Au/Co/Au nanosandwiches with enhanced magneto-optical activity. *Small* **2008**, *4*, 202–205.

10. Jain, P.K.; Xiao, Y.; Walsworth, R.; Cohen, A.E. Surface plasmon resonance enhanced magneto-optics (SuPREMO): Faraday rotation enhancement in gold-coated iron oxide nanocrystals. *Nano Lett.* **2009**, *9*, 1644–1650.

11. González-Díaz, J.B.; Sepúlveda, B.; García-Martín, A.; Armelles, G. Cobalt dependence of the magneto-optical response in magnetoplasmonic nanodisks. *Appl. Phys. Lett.* **2010**, *97*, 043114.

12. Dani, R.K.; Wanh, H.; Bossmann, S.H.; Wysin, G.; Chikan, V. Faraday rotation enhancement of gold coated Fe2O3 nanoparticles: Comparison of experiment and theory. *J. Chem. Phys.* **2011**, *135*, 224502.

13. Pakdel, S.; Miri, M. Faraday rotation and circular dichroism spectra of gold and silver nanoparticle aggregates. *Phys. Rev. B* **2012**, *86*, 235445.

14. Armelles, G.; Cebollada, A.; García-Martín, A. ; Gonzá M.U. Magnetoplasmonics: Combining magnetic and plasmonic functionalities. *Adv. Opt. Mater.* **2013**, *1*, 10–35.

15. Christofi, A.; Stefanou, N. Nonreciprocal optical response of helical periodic structures of plasma spheres in a static magnetic field. *Phys. Rev. B.* **2013**, *87*, 115125.

16. Christofi, A.; Stefanou, N. Nonreciprocal photonic surface states in periodic structures of magnetized plasma nanospheres. *Phys. Rev. B.* **2013**, *88*, 125133.

17. Nahal, A.; Talebi, R. Ellipticity-dependent laser-induced optical gyrotropy in AgCl thin films doped by silver nanoparticles. *J. Nanopart. Res.* **2014**, *16*.

18. Soljacˇic´, M.; Luo, C.; Joannopoulos, J.D.; Fan, S. Nonlinear photonic crystal microdevices for optical integration. *Opt. Lett.* **2003**, *28*, 637–639.

19. Scalora, M.; Dowling, J.P.; Bowden, C.M.; Bloemer, M.J. Optical limiting and switching of ultrashort pulses in nonlinear photonic band gap materials. *Phys. Rev. Lett* **1994**, *73*, 1368–1371.

20. Mingaleev, S.F.; Kivshar, Y.S. Nonlinear transmission and light localization in photonic-crystal waveguides. *J. Opt. Soc. Am. B* **2002**, *19*, 2241–2249.

21. Yu, Z.; Fan, S. Complete optical isolation created by indirect interband photonic transitions. *Nat. Photon.* **2009**, *3*, 91–94.

22. Hwang, J.; Song, M.H.; Park, B.; Nishimura, S.; Toyooka, T.; Wu, J.W.; Takanishi, Y.; Ishikawa, K.; Takezoe, H. Electro-tunable optical diode based on photonic bandgap liquid-crystal heterojunctions. *Nat. Mater.* **2005**, *4*, 383–387.

23. Lu, C.; Hu, X.; Zhang, Y.; Li, Z.; Xu, X.; Yang, H.; Gong, Q. Ultralow power all-optical diode in photonic crystal heterostructures with broken spatial inversion symmetry. *Appl. Phys. Lett.* **2011**, *99*, 051107.

24. Wang, C.; Zhou, C.Z.; Li, Z.Y. On-chip optical diode based on silicon photonic crystal heterojunctions. *Opt. Express***2011**, *19*, 26948–26955.

25. Serebryannikov, A.E.; Nojima, S.; Ozbay, E. One-way absorption of terahertz waves in rod-type and multilayer structures containing polar dielectrics. *Phys. Rev. B* **2014**, *90*, 235126.

26. Lockyear, M.J.; Hibbins, A.P.; White, K.R.; Sambles, J.R. One-way diffraction grating. *Phys. Rev. E* **2006**, *74*, 056611.

27. Serebryannikov, A.E. One-way diffraction effects in photonic crystal gratings made of isotropic materials. *Phys. Rev. B***2009**, *80*, 155117.

28. Kuzmiak, V.; Maradudin, A.A. Asymmetric transmission of surface plasmon polaritons. *Phys. Rev. A* **2012**, *86*, 043805.

29. Yannopapas, V. One-way photonic band gaps and optical isolation with three-dimensional photonic crystals of low symmetry. *Phys. Rev. A* **2013**, *88*, 043837.

30. Stefanou, N.; Karathanos, V.; Modinos, A. Scattering of electromagnetic waves by periodic structures. *J. Phys. Condens. Matter* **1992**, *4*, 7389–7400.

31. Stefanou, N.; Yannopapas, V.; Modinos, A. Heterostructures of photonic crystals: Frequency bands and transmission coefficients. *Comput. Phys. Commun.* **1998**, *113*, 49–77.

32. Stefanou, N.; Yannopapas, V.; Modinos, A. MULTEM 2: A new version of the program for transmission and band-structure calculations of photonic crystals. *Comput. Phys. Commun.* **2000**, *132*, 189–196.

33. Gantzounis, G.; Stefanou, N.; Papanikolaou, N. Optical properties of periodic structures of metallic nanodisks. *Phys. Rev. B* **2008**, *77*, 035101.

34. Yannopapas, V.; Vanakaras, A.G. Layer-multiple-scattering theory for metamaterials made from clusters of nanoparticles. *Phys. Rev. B* **2011**, *84*, 085119.

35. Yannopapas, V. Layer-multiple-scattering method for photonic structures of general scatterers based on a discrete-dipole approximation/T-matrix point-matching method. *J. Opt. Soc. Am. B* **2014**, *31*, 631–636.

36. Kim, S.H.; Lee, S.Y.; Yang, S.M.; Yi, G.R. Self-assembled colloidal structures for photonics. *NPG Asia Mater.* **2011**, *3*, 25–33.

37. Ishii, M.; Kato, H.; Hashimoto, I.; Homma, Y. Synthesis of sapphire nanoparticles with graphite shells by hot-filament

chemical vapor deposition. *Mater. Express* **2014**, *4*, 135–143.

38. Istrate, E.; Sargent, E.H. Photonic crystal heterostructures and interfaces. *Rev. Mod. Phys.* **2006**, *78*, 455–481.

39. Kedia, S.; Reddy, M.S.; Vijaya, R. Photonic crystal based direct and inverse heterostructures by colloidal self-assembly.*Opt. Photon. J.* **2012**, *2*, 242–248.

40. Cai, Z.; Liu, Y.J.; Teng, J.; Lu, X. Fabrication of large domain crack-free colloidal crystal heterostructures with superposition bandgaps using hydrophobic polystyrene spheres. *ACS Appl. Mater. Interface* **2012**, *4*, 5562–5569.

41. Liu, X.; Zhao, D.; Geng, C.; Zhang, L.; Tan, T.; Hu, M.; Yan, Q. Fabrication of colloidal photonic crystal heterostructures free of interface imperfection based on solvent vapor annealing. *J. Colloid Interface Sci.* 2014, *434*, 98–103.

CHAPTER 4

Lamb Waves Propagation along 3C-SiC/AlN Membranes for Application in Temperature-Compensated, High-Sensitivity Gravimetric Sensors

Cinzia Caliendo [1,*], Arnaldo D'Amico [2] and Fabio Lo Castro [3]

[1] Istituto dei Sistemi Complessi, ISC-CNR, Via Salaria Km. 29.300, Monterotondo Scalo, 00015 Roma, Italy

[2] Dipartimento di Ingegneria Elettronica, Università di Tor Vergata, Via del Politecnico 1, 00133 Roma, Italy

[3] Istituto di Acustica e Sensoristica "Orso Mario Corbino", IDASC-CNR, Via del Fosso del Cavaliere 100, 00133 Roma, Italy

ABSTRACT

The propagation of the fundamental quasi-symmetric Lamb mode S_0 travelling along 3C-SiC/*c*-AlN composite plates is theoretically studied with respect to the AlN and SiC film thickness, the acoustic wave propagation direction and the electrical boundary conditions. The temperature effects on the phase velocity have been considered for four AlN/SiC-based electroacoustic coupling configurations, specifically addressing the design of temperature-compensated, enhanced-coupling, GHz-range electroacoustic devices. The gravimetric sensitivity and resolution of the four temperature-stable SiC/AlN composite structures are theoretically investigated with respect to both the AlN and SiC sensing surface. The SiC/AlN-based sensor performances are compared to those of surface acoustic waves and Lamb S_0mode mass sensors implemented on bulk conventional piezoelectric materials and on thin suspended membranes.

KEYWORDS

AlN; SiC; suspended membranes; S_0 Lamb mode; thermal stability; sensors

1. INTRODUCTION

In recent years AlN piezoelectric thin films have been used to fabricate a wide variety of radio frequency resonators and filters for applications in telecommunication and sensing fields, thanks to their high acoustic wave velocity, good electromechanical coupling coefficient (K^2), and resistance to severe environmental conditions [1]. Cubic polytype SiC (3C-SiC) substrates have been proven to be suitable for the implementation of AlN-based multilayered devices thanks to some interesting properties such as low mechanical loss, a high surface acoustic wave (SAW) velocity, resistance to chemicals, high hardness, the low lattice mismatch (1%) and a thermal expansion coefficient that closely matches that of AlN. Moreover AlN and SiC show opposite temperature coefficients of delay (TCD), thus allowing the fabrication of temperature-compensated electroacoustic devices for the proper

AlN and SiC films thickness. Large-area low-cost 3C-SiC(100) and (111) films on Si(001) and (111) substrates are commercially available [2,3] for limited thickness ranges (up to 1 and 20 μm thick). Highly *c*-axis oriented AlN films have been grown on 3C-SiC films by using different techniques, such as metalorganic vapor phase epitaxy (MOVPE) [4], low pressure metalorganic chemical vapor deposition (LP-MOCVD) [5], and reactive magnetron sputtering [6]. The latter technique can also be employed to grow metal films (such as Pt or Mo) to be patterned for the implementation of interdigital transducers (IDTs) on the AlN free surface or at the SiC/AlN interface. The Si/SiC/AlN-based structures can take advantage from the anisotropic bulk Si micromachining technique to etch away the silicon lying under the SiC film and leave the SiC/AlN/IDTs as a freely suspended membrane.

Recently, the fundamental quasi-symmetric Lamb wave mode S_0 propagating along 3C-SiC/c-AlN thin membranes has been theoretically investigated, confirming the *c*-axis oriented AlN thin films on epitaxial 3C-SiC layers as an enabling technology for high-frequency and high-*Q* Lamb wave devices [7]. In the present study the effect of the temperature on the propagation of the S_0 Lamb mode along 3C-SiC/c-AlN composite plates is studied. Our goal is to provide the design parameters required for gravimetric sensors able to work in harsh environments and showing remarkable performance features such as high-frequency operation, applicability of temperature-compensation techniques, and enhanced-K^2.

2. THEORETICAL INVESTIGATION OF THE ACOUSTIC WAVES PROPAGATION IN 3C-SIC/C-ALN

In 3C-SiC/c-AlN-based electroacoustic devices the anisotropy of the SiC layer strongly affects the characteristics of the acoustic waves' (AWs) propagation rather than the c-AlN that is isotropic in the c plane. Hence the Lamb modes propagation characteristics are influenced by factors other than by the AlN layer thickness such as the SiC thickness, cut and propagation direction. The AWs' propagation along 3C-SiC/c-AlN layered substrates was investigated by exploiting the numerical code developed by Adler and coworkers at McGill University (Montreal, Canada): a detailed description of the propagation of surface and plate acoustic waves in layered structures can be found in Farnell and Adler's work [8]. The velocity calculations were performed under the hypothesis of lossless materials, and assuming the single crystal AlN and 3C-SiC material constants available in the literature [9,10]. The SiC/AlN composite plate was considered to be an infinite plate in the x and $z$directions, being x the AW propagation direction.

2.1. Lamb Modes

Two sets of waves propagate along a 3C-SiC plate with finite thickness: the symmetric Lamb waves, whose particle displacements are symmetric about the neutral plane of the plate, and the antisymmetric Lamb waves, whose displacements have odd symmetry about the neutral plane. For sufficiently thin plate only two waves exist: the fundamental symmetric and antisymmetric modes, S_0 and A_0[11]. The A_0 mode is easily identified by its reducing velocity as the plate thickness approaches zero. The S_0 mode has a weak dispersion region, for small plate thicknesses, where its velocity approaches vs2√where v_s is the shear bulk wave in an unbounded propagating medium. With increasing the plate thickness, the two modes asymptotically reach the SAW velocity in the substrate. Higher order modes exist only starting from a minimum plate thickness value (the cut off) and their velocity asymptotically reach the slow bulk shear velocity in the substrate with increasing plate thickness. A *composite plate* consisting of two different materials, *i.e.*, 3C-SiC/c-AlN plate, has no symmetry about the central plane and the modes are no longer rigorously separated into symmetric and antisymmetric sets. It is intended that the two fundamental modes, arbitrary cited S_0 and A_0 modes in the present paper, have to be understood as*quasi-symmetric* and *quasi-antysymmetric* modes [12].

Figure 1(a,b) show, as an example, the slowness curve of the S_0 and A_0 modes propagating along 3C-SiC/c-AlN composite plate for SiC thickness-to-wavelength-ratio h/λ_{SiC} = 0.1 and AlN normalized thicknesses h/λ_{AlN} = 0, 0.001, 0.05 and 0.1.

For both the <100> and <110> directions of the (001) plane of 3C-SiC, the velocity of the two modes asymptotically reaches the velocity of the SAW in the piezoelectric layer with increasing AlN normalized thickness, h/λ_{AlN}. With increasing the SiC layer thickness, the velocities of the two fundamental modes asymptotically converge to the velocity of the first Rayleigh-like mode propagating along these two structures. For propagation directions different from 0° and 45°off x axis in the (001) plane, the phase and group velocity of the propagating modes are not collinear and the direction of energy transport deviates from the phase velocity vector by an angle, the power flow angle (PFA), that depends on the substrate anisotropy and on the deviation angle ϑ with respect to the selected wave vector direction ϑ_0. Referring to Figure 1(a,b), the phase velocity vector is the radius-vector of the curve drawn from the origin: it determines the wave propagation direction. The direction of the group velocity vector is normal to the curve at the point of intersection of the curve with the wave propagation vector. As it can be seen in Figure 1(a,b), the phase and group velocities are thus collinear for $\vartheta_0 = 0°$ and 45°: for deviation from these directions, the PFA is not null and the resulting beam steering vanishes the practical interest of the device [13,14].

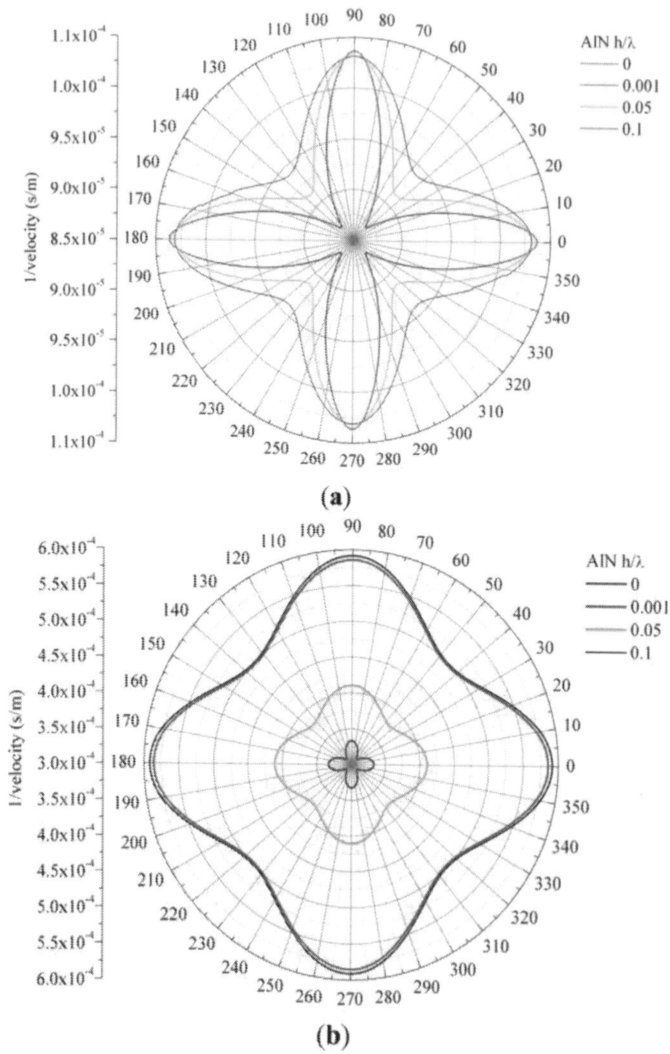

Figure 1. Slowness curves. (**a**) Slowness curves of the S_0 mode propagating along 3C-SiC plate (0.1 h/λ thick) covered by an AlN layer; (**b**) slowness curves of the A_0 mode propagating along 3C-SiC plate (0.1 h/λ thick) covered by an AlN layer. In both cases the AlN h/λ (0, 0.001, 0.05 and 0.1) is the running parameter.

Figure 2(a,b) show the calculated phase velocity dispersion curves of the S_0 mode propagating along the c-AlN/SiC(001)<100> and <110>, for different h/λ_{SiC} values. As it can be seen from Figure 2, higher phase velocities can be reached for AWs propagating along the <110> direction with respect to the waves propagating along <100>. For $h/\lambda_{AlN} \ll 1$, the velocity decreases as the SiC thickness increases; for the same h/λ_{SiC} value, the velocity decreases with increasing the h/λ_{AlN} value. For large AlN and SiC thicknesses, the top and bottom surfaces of the plate become essentially decoupled and the velocity in the plate decreases, matching that of the *Rayleigh-like* mode in AlN/SiC.

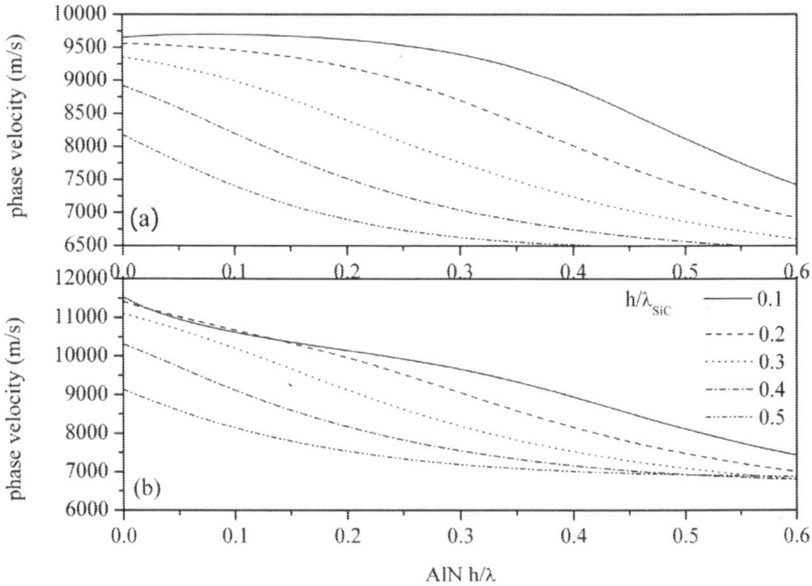

Figure 2. The dispersion curves for the S_0 Lamb mode propagating (**a**) along c-AlN/SiC(001)<100> and (**b**) along c-AlN/3C-SiC(001)<110>.

The weak dispersion and the high velocity of the S_0 mode (for $h/\lambda_{AlN} \ll 1$) reduces the AlN thickness uncertainty and enables GHz range (from 6 to 10 GHz) resonators with IDTs' line-width resolution from 0.4 to 0.25 μm.

2.2. The Electroacoustic Coupling Coefficient, K^2

The performances of an electroacoustic device depend on the materials chosen and on the transduction configuration adopted. To find the optimum device configuration, the characteristics of the Lamb mode propagation and transduction are studied as a function of the SiC and AlN layer thickness, the ground electrode and IDTs location with respect to the AlN film surface.

An ideal electric conductor, placed at one or both the surfaces of the piezoelectric layer, shorts the tangential component of the electric field, thus affecting its distribution into the layer, and resulting in a decrease in the acoustic phase velocity. The electroacoustic coupling coefficient, K^2, is based on the computation of the difference between the phase velocity of the AW propagating along a free piezoelectric surface and that along a metalized piezoelectric surface. Only the electric loading from the IDTs and ground electrode is taken into account, omitting the mass loading effect due to their finite thickness: the IDT metal pattern, as well as the ground electrode, is considered ideal, *i.e.*, perfectly conducting and infinitely thin.

The electroacoustic coupling coefficient, K^2, is a parameter of primary importance for the design of AW devices as only $K^2$of the total input of the electrical energy is converted to the elastic energy of the AW. For AW propagating along layered structures, the K^2 is frequency dispersive and depends on the type and orientation of the piezoelectric material, on the propagation direction and on the electrical boundary conditions. K^2 can be approximated as:

$$K^2 = 2 \cdot \frac{v_{ph}^f - v_{ph}^m}{v_{ph}^f} \qquad (1)$$

where v_{ph}^f and v_{ph}^m are the phase velocities along the electrically open and shorted surfaces of the AlN film. The vmph is obtained by the insertion of a perfectly conductive and infinitesimally thin metal film at the interfaces where the IDTs and the ground plane are located.

In the AlN/SiC composite plate, four piezoelectric coupling configurations can be obtained by placing the IDT's at the substrate/film interface (Substrate/Transducer/Film, STF), at the film surface (Substrate/Film/Transducer, SFT), farther including a ground electrode opposite the IDTs (Substrate/ Transducer/ Film/Metal and Substrate/Metal/Film/ Transducer, STFM and SMFT). The four coupling configurations are shown in Figure 3.

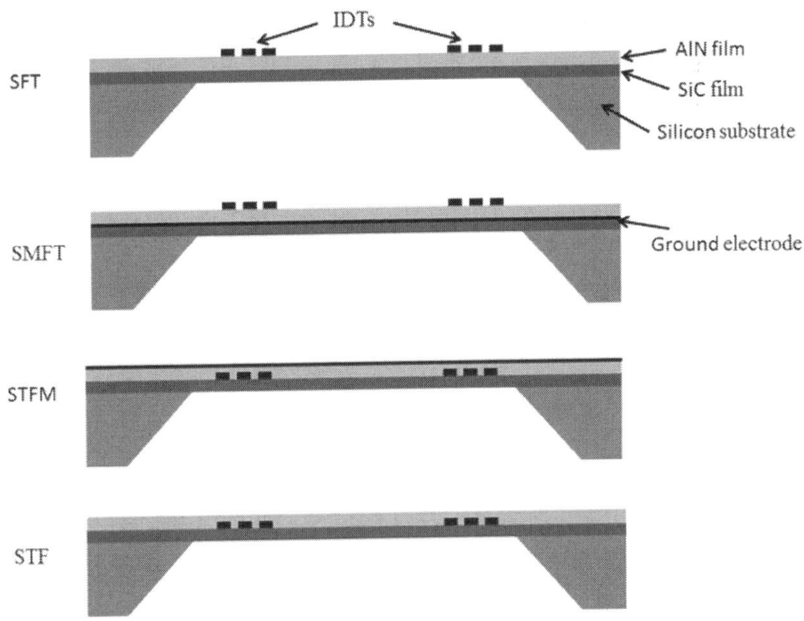

Figure 3. The four coupling configurations.

The K^2 dispersion curves of the four configurations were evaluated by the relations $K^2_{SFT} = 2 \cdot \frac{v_{ph}^{SF} - v_{ph}^{SFM}}{v_{ph}^{SF}}$, $K^2_{SMFT} = 2 \cdot \frac{v_{ph}^{SMF} - v_{ph}^{SMFM}}{v_{ph}^{SMF}}$, $K^2_{STF}2 = 2 \cdot \frac{v_{ph}^{SF} - v_{ph}^{SMF}}{v_{ph}^{SF}}$ and $K^2_{STFM} = 2 \cdot \frac{v_{ph}^{SFM} - v_{ph}^{SMFM}}{v_{ph}^{SFM}}$ for the SFT, SMFT, STF and STFM configurations, respectively, being the SiC and the AlN layers represented as the Substrate and the Film, respectively. The K^2 dispersion curves (K^2 vs h/λ_{AlN}) of each AlN/SiC-based configuration were calculated for different h/λ_{SiC} values and for the <100> and <110> propagation directions. The calculated data addressing the AlN/SiC(001)<100> and AlN/SiC(001)<110> are shown in Figures 4 and 5, being the h/λ_{SiC} the running parameter.

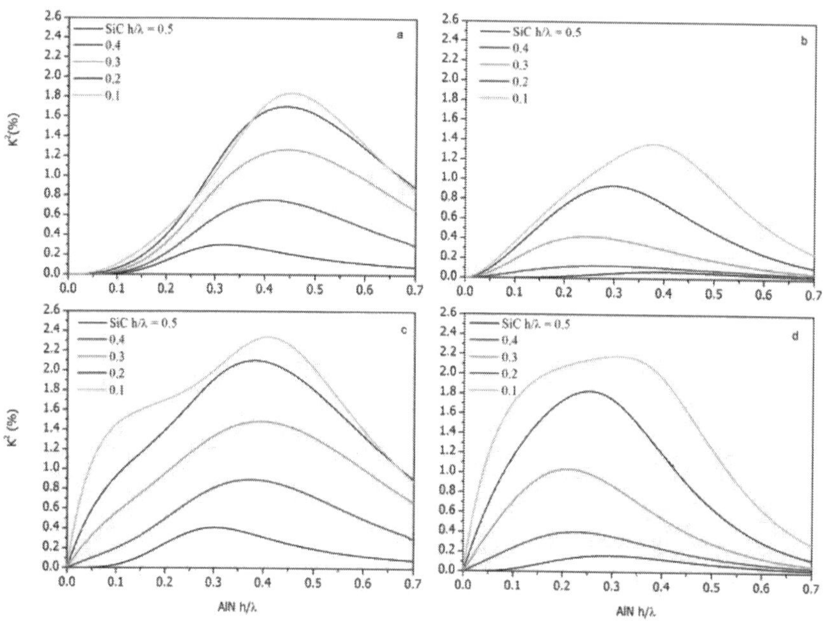

Figure 4. The K^2 dispersion curves of the (**a**) STF, (**b**) SFT, (**c**) STFM, and (**d**) SMFT coupling configurations on c-AlN/SiC(001)<100> substrates; the h/λ_{SiC} (from 0.1 to 0.5) is the running parameter.

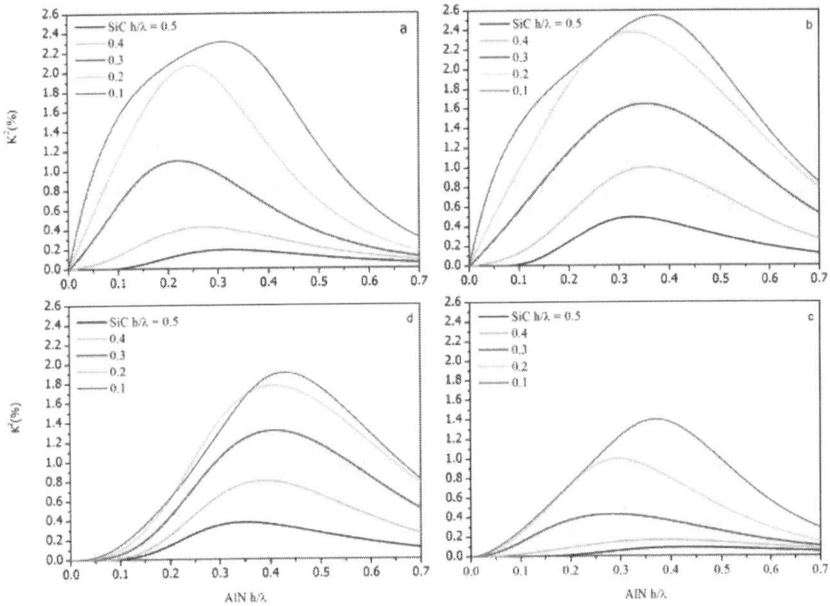

Figure 5. The K^2 dispersion curves of the **(a)** SMFT, **(b)** STFM, **(c)** SFT, and **(d)** STF coupling configurations on c-AlN/SiC(001)<110> substrates; the h/λ_{SiC} (from 0.1 to 0.5) is the running parameter.

For acoustic waves propagating along layered structures the achievable K^2 can be larger than that of the individual piezoelectric material (K^2 = 0.28% for AlN). For surface acoustic waves (SAWs) propagating along the free surface of the piezoelectric semi-infinite substrate the energy localization occurs within a thickness of a wavelength, while the Lamb wave's energy is predominantly confined in the piezoelectric layer whose thickness is less than a wavelength. Moreover, the K^2 dispersion curve is strongly affected by the electrical boundary conditions that influence the potential depth profile: K^2reaches high values when a strong electric field is induced in the piezoelectric layer, and it decreases as the thickness of the SiC layer increases. The STFM and SMFT structures exhibit the highest K^2 value due to the strong electric field that can be induced between the IDT and the electrode.

2.3. Temperature Coefficient of Delay

The velocity of the AWs is a function of the material constants (mass density, elastic, piezoelectric and dielectric constants) of the propagating medium, and these constants are affected by the temperature changes: as a result the AW velocity is sensitive to the temperature variations of the surroundings. The S_0 mode phase velocity at different temperatures was theoretically estimated by modifying the mass density and the elastic constants of AlN and SiC according to their temperature coefficients available from the literature. Table 1 lists the SiC and AlN coefficients of thermal expansion (CTE) and the temperature coefficients of elastic constants used in our calculations.

Table 1. The SiC and AlN CTE and the temperature coefficients of the elastic constants used in the numerical calculations.

Material	Tcc_{11}^{E} (ppm/°C)	Tcc_{33}^{E} (ppm/°C)	Tcc_{12}^{E} (ppm/°C)	Tcc_{13}^{E} (ppm/°C)	Tcc_{44}^{E} (ppm/°C)	Ref.	CTE (ppm/°C)	Ref.
AlN	−80	−100	−180	−160	−50	9	//c 5.27 ⊥c 4.15	[9]
3C-SiC	71	-	78	-	30	15, 16	3.8	[15]

As it can be seen AlN and SiC have opposite sign temperature coefficients of the elastic constants [9,15–21]. The AlN and SiC mass density and elastic constants were evaluated at different temperatures T by modifying their values according to their temperature coefficients in the following manner: $\rho(T) = \rho(20\ °C)[1 - (\alpha_{11} + \alpha_{22} + \alpha_{33})\Delta T]$ where $\rho(T)$ is the mass density at a certain temperature T, $\Delta T = T-20$ °C, $\rho(20\ °C)$ is the mass density at 20 °C and α_{ii} is the linear temperature expansion coefficient (available in the literature). The elastic constants c_{ij} were calculated at different temperatures as follows: $c_{ij}(T) = c_{ij}(20\ °C)(1 + Tcc_{ij}\Delta T)$, where $c_{ij}(20\ °C)$ and $c_{ij}(T)$ represent the elastic constants at 20 °C and at a certain T, Tcc_{ij} is the temperature coefficient of the elastic constant c_{ij}[22]. The temperature coefficient of velocity (TCV) of the four coupling configurations was assessed through the fractional change of the acoustic velocity, $[(v_{T°C} - v_{20°C})/v_{20°C}]$, as a function of T, being $v_{20°C}$ and $v_{T°C}$ the S_0 mode phase velocity at ambient temperature and at a certain temperature T, for T ranging from −20 to 200 °C. The theoretical $TCD = \alpha_{eff} - TCV$ of the four configurations was calculated for different $h/\lambda_{AlN}$and h/λ_{SiC} values, being α_{eff} the effective thermal expansion coefficient of the composite plate along the propagation direction, computed applying the model for thermal bimorph actuators [23]. A temperature compensated structure is obtained when the temperature coefficient of frequency $TCf = (1/f)(\partial)f/(\partial)T = -TCD$ is zero. The temperature compensated points (TCPs; are the AlN normalized thicknesses, h/λ_{AlN}, at which α_{eff} equilibrates the TCV to form a thermally compensated structure with the first order TCD equal to 0 ppm/°C. Table 2 lists the following parameters: the TCP of each configuration, the K^2 and the phase velocity of the thermally compensated configurations on c-AlN/3C-SiC(001)<100> and <110> structures.

The listed data are relative to $h/\lambda_{SiC} = 0.1$ and 0.2 only, since at higher SiC normalized thicknesses the K^2 of the corresponding TCPs rapidly decreases.

2.4. Gravimetric Sensitivity

The most common sensing application of the electroacoustic devices is based on the gravimetric principle for mass detection. A mass accumulation on the device surface changes the surface density of the propagating medium, hence resulting in an AW velocity shift. If the added mass consists of an ideal thin elastic film that moves synchronously with the oscillating surface, the fractional-velocity-change to added-mass ratio defines the sensor's gravimetric sensitivity $S_m = [(v_l - v)/v_l]/m$, being $m = \rho h$, ρ and h the added layer's mass density and thickness, v and v_l the unloaded and mass-loaded plate's phase velocity. For a resonator or a delay-line oscillator sensor, the sensitivity can be expressed as $S_m = [(f - f_0)/f]/m$, being f and f_0 the perturbed and unperturbed resonant frequency. The theoretical gravimetric sensitivity S_m of the thermally compensated AlN/SiC structures was deduced by numerical computer calculation [8] of the S_0 mode velocity along the plate mass-loaded by a thin lossless SiO_2 overlay, for different SiO_2 surface mass density values $\rho_{SiO2} \cdot h/\lambda_{SiO2}$ (ρ_{SiO2} = 2,200 Kg/m^3). Since the gravimetric detection can occur either on the AlN or SiC plate surface, two normalized gravimetric sensitivity values S_m/λ were calculated for each configuration. Table 3 lists the sensitivity of the AlN and SiC sensing surfaces of the SiC/AlN composite plate.

These sensitivity values were deduced by the calculation of the relative velocity shift per unit mass loaded for each configuration: the unloaded and mass loaded velocity values, v and v_l, were calculated taking into account the electric loading from the IDTs and the ground electrode. The mass loading effect from the IDTs and the ground electrode was not taken into account.

From Table 3 it appears that the gravimetric detection occurring at the AlN or SiC sensing surface of the same plate shows different efficiency. In a composite plate consisting of two different materials, such as AlN/SiC, the AW energy confinement is different close to one or the other plate's surface: the highest gravimetric sensitivity is achieved for the plate sensing surface that transmits an higher amount of acoustic energy to the mass loading layer [24–26]. Figure 6 shows, as an example, the normalized longitudinal and shear vertical particle displacement components U1/UAlN surface1and U3/UAlN surface1 *vs*the penetration depth across the SiO_2/AlN- and SiO_2/SiC- interfaces of the c-AlN/ SiC(001)<110> plate, being zero the shear horizontal displacement component U_2. Figure 6 refers to the SFT configuration loaded by a lossless SiO_2 layer 0.01 h/λ thick, for $h/\lambda_{SiC} = 0.1$. From Figure 6 it results that the magnitude of the displacement components at the AlN surface is larger than that at the SiC surface; consequently the relative fraction of the energy density stored in the layer added at the AlN surface is higher than that at the SiC surface, in accordance with the theoretically predicted sensitivity.

Table 2. The TCP of each configuration, the K^2 and phase velocity of the thermally compensated configurations.

Structure	h/λ_{SiC}	Config.	TCP	K^2_{TCP} (%)	Velocity (m/s)
c-AlN/SiC(001)<100>	0.1	SFT	0.334	1.31	9,235
		SMFT	0.324	2.16	9,200
		STF	0.331	1.26	9,250
		STFM	0.327	2.11	9,187
	0.2	SFT	0.464	0.55	7,536
		SMFT	0.463	0.83	7,527
		STF	0.473	1.68	7.514
		STFM	0.467	1.93	7.524
c-AlN/SiC(001)<110>	0.1	SFT	0.236	0.94	9,980
		SMFT	0.229	2.17	9,943
		STF	0.233	0.85	9,995
		STFM	0.233	2.11	9,923
	0.2	SFT	0.413	0.76	8,058
		SMFT	0.401	1.25	8.070
		STF	0.413	1.77	8.100
		STFM	0.403	2.22	8.075

Both sides of the four configurations sense the mass and the acoustoelectric loadings from the added layer, except the metalized

AlN top surface of the STFM configuration and the SiC bottom surface of the SMFT configuration which can sense only the gravimetric perturbation, as a consequence of the presence of the ground electrode that shorts the electric field associated to the acoustic wave. The acoustoelectric loading produces a decrease in phase velocity and also an increased attenuation for a specific sheet resistance range of the added layer [27]. The fractional phase velocity change induced by the acoustoelectric effect can be isolated from the mass sensor response by subtracting the mass accumulation effect measured by a reference mass sensor [27], such as the SiC-surface of the SMFT or the metalized AlN surface of the STFM.

Table 3. The gravimetric sensitivity S_m and the limit of mass resolution R_m of the temperature compensated coupling configurations loaded on the AlN or SiC surface with a SiO_2 film; λ(μm) is the numerical value of the acoustic wavelength expressed in μm.

Structure	SiC (h/λ) 0.1		0.2	
	[a] $-S_m/\lambda$(μm) $(-cm^2 \cdot g^{-1})$	[b] $R_m \cdot \lambda^2$ (μm) $(10^{-13}\ g \cdot cm^{-2})$	[c] $-S_m/\lambda$(μm) $(-cm^2 \cdot g^{-1})$	[b] $R_m \cdot \lambda^2$ (μm) $(10^{-13}\ g \cdot cm^{-2})$
c-AlN/SiC(001)<100>				
SFT	2,306; 2,770	1.41; 1.17	2,859; 3,609	1.39; 1.10
SMFT	2,384; 2,714	1.37; 1.20	2,804; 3,598	1.42; 1.11
STF	2,327; 2,775	1.39; 1.17	2,764; 3,624	1.44; 1.10
STFM	2,329; 2,979	1.40; 1.10	2,874; 3,395	1.39; 1.17
c-AlN/SiC(001)<110>				
SFT	3,693; 2,385	0.81; 1.26	3,160; 3,588	1.18; 1.04
SMFT	3,847; 2,351	0.78; 1.28	3,166; 3,567	1.17; 1.04
STF	3,808; 2,380	0.79; 1.27	3,084; 3,640	1.20; 1.02
STFM	3,718; 2,347	0.82; 1.29	3,234; 3,520	1.15; 1.06

[a] for SiO_2 h/λ = 0 ÷ 0.04. [b] $f_0 = v/\lambda$, being v the velocity of the temperature compensated configuration.
[c] for SiO_2 h/λ = 0 ÷ 0.1.

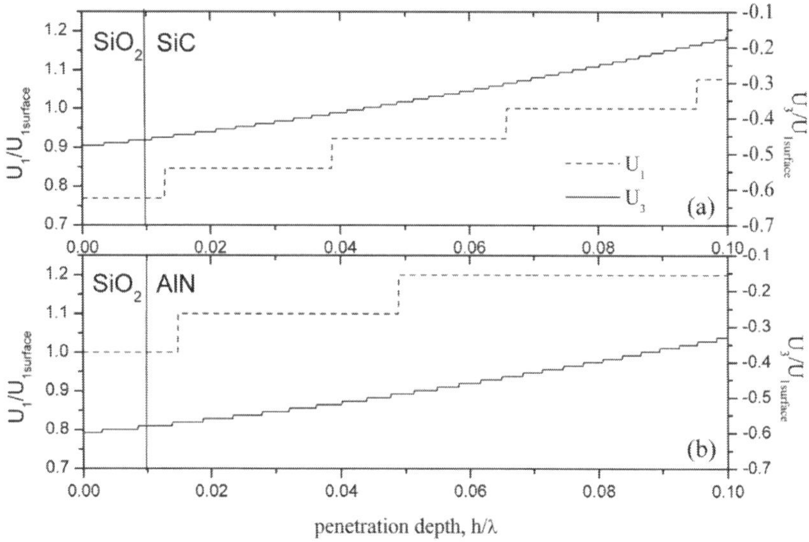

Figure 6. The normalized particle displacement components *vs.* the penetration depth (calculated by 0.0001 h/λ steps) along the (**a**) SiO_2/SiC and (**b**) SiO_2/AlN interfaces of the AlN/SiC(001)<110>-SFT plate.

The mass resolution R_m of a gravimetric sensor is defined as the smallest detectable surface mass per unit area. For a resonator- or for a delay-line oscillator-based sensor, R_m is set by the resonator stability-to-sensitivity ratio:

$$R_m = \lim_{\Delta f \to \Delta f_{noise}} \frac{(\Delta f / f_0)}{S_m} \tag{2}$$

where $\Delta f = f - f_0$ is the output response, and Δf_{noise} is defined as three times the frequency random fluctuation, $\Delta f = 3\Delta f_{noise}$[27]. Temperature-compensated sensors are useful in minimizing the effects of the temperature on the oscillation frequency of the device. R_m has been calculated under the hypothesis of $\Delta f_{noise} = 1$ Hz, at $f_0 = v_{TCP}/\lambda$, being v_{TCP} the phase velocity of the S_0 mode evaluated at the TCPs; the R_m data are listed in Table 3. As a rule,

S_m can be increased by reducing the suspended membrane thickness, but in practical cases the lower limit of the c-AlN and 3C-SiC films thickness is set by the structural properties of the layers. Extremely thin AlN films can show very low K^2 as a consequence of the poor c-axis orientation of the grains, and extremely thin SiC films on Si substrates could be amorphous and lead to degraded device performances. As an example, for $\lambda = 10$ μm, the zero-TCD configurations will show h/λ_{SiC} from 1 to 2 μm, h/λ_{AlN} from ~2 to ~5 μm, S_m from ~230 to ~385 g/cm^2, R_m from 8 to 14 pg/cm^2, f_0 from ~750 to ~900 MHz, and K^2 from 0.55 to 2.22%.

The gravimetric sensor can also be used for the characterization of film properties such as film thickness and surface area; it is also useful to monitor processes such as thin film deposition or removal, and materials modifications [27]. The mass resolution values R_m listed in Table 3 are referred to 1 Hz of frequency fluctuation, as it is reasonable in ST quartz resonators, just to cite an example. For a thin film device, the frequency fluctuation is higher as a consequence of the polycrystalline nature of the films although highly oriented. Supposing to apply the SiC/AlN-based sensor to monitor the deposition of a thin SiO_2 layer, a film thickness resolution in the range from 1 to 2×10^{-12} m can be obtained under the following hypothesis: $\lambda = 2$ μm and TCD = 0.5 ppm/°C; in this case the frequency fluctuation is few hundreds of Hz. A lower thickness resolution can be obtained by further reducing the acoustic wavelength.

As shown in Table 3, the AlN sensing side of the structures implemented on SiC(001)<110>/AlN (for $h/\lambda_{SiC} = 0.1$) seem to be the most promising as they correspond to the highest sensitivity and lowest resolution values. The SiC sensing side is equally highly sensitive for both the two substrates with $h/\lambda_{SiC} = 0.2$. Among the four configurations, the SFT and SMFT are the most promising as, although they show moderate K^2, they offer the advantage of a robust technology over the STF and STFM structures for which the IDT strips under the AlN layer can result in thin films cracks. The AlN crystalline quality affects the piezoelectric characteristics of the film and thus of the device performances; thus high-quality AlN film is crucial to achieve temperature compensated structures.

In reference [28] the S_m and R_m of *SAW* sensors based on both conventional piezoelectric substrates (such as ST-x quartz, yx-, yz-, and yz- 128.086° rot. x-$LiNbO_3$) and AlN film on a semiinfinite 3C-SiC substrates were assessed through the fractional change of the acoustic velocity as a function of the mass loading layer thickness. At f_0 = 100 MHz, the former substrates exhibit good S_m (from 62 to 113 cm^2/g) and remarkable R_m (from 0.27 to 0.45 ng/cm^2), but unfortunately $LiNbO_3$suffers from a very large TCD while the ST-quartz shows a very small K^2. Moreover the quartz and $LiNbO_3$ substrates are incompatible with monolithic integration technology, and suffer limited *SAW* velocity (from ~3,000 to ~4,000 m/s). The AlN/SiC-based *SAW* configurations, for f_0 = 100 MHz, show the following properties: S_m ~70 cm^2/g, R_m ~0.4 ng/cm^2, $K^2$from 0.71 to 1.18%, TCD from ~−10 to 5 ppm/°, and phase velocity from ~5,700 to ~6,800 m/s. The AlN/SiC-based S_0 mode configurations show the following properties for f_0 = 100 MHz: S_m from 25 to 52 cm^2/g, R_m from 0.58 to 1.2 ng/cm^2, K^2 from 0.55 to 2.22%, TCD = 0 ppm/°C, and the phase velocity from 7,500 to 10,000 m/s. For the same operation frequency of 100 MHz, the *SAW* sensors are more sensitive than the temperature-compensated S_0 mode sensors, but their K^2, phase velocity and TCD are less attractive.

In reference [29] the S_0 mode mass sensor implemented on a GaN suspended membrane is theoretically and experimentally investigated: for λ = 16 μm, the sensor operating frequency is 471.3 MHz and its sensitivity is 174 cm^2/g. For the same wavelength, the AlN/SiC(000) and (0045) sensitivity ranges from 143 to 244 cm^2/g (for h/λ_{SiC} = 0.1) and from 181 to 256 cm^2/g (for

$h/\lambda_{SiC} = 0.2$), while the operating frequency ranges from 577 to 621 MHz, and from 471 to 506 MHz, respectively.

In reference [30] the S_0 mode mass sensitivity (equal to 557 cm^2/g) of a suspended AlN membrane, 2 μm thick, coated with a hexamethyldisiloxane (HMDSO)-plasma-polymerized thin film (75 nm thick, and showing a mass density of 1,100 Kg/m^3), was estimated for $\lambda = 12$ μm and 888 MHz operating frequency. For the same wavelength, the SiC/AlN-sensor sensitivity ranges from 190 to 325 cm^2/g (for $h/\lambda_{SiC} = 0.1$) and from 242 to 341 cm^2/g (for $h/\lambda_{SiC} = 0.2$), while the operating frequency ranges from 770 to 829 MHz, and from 628 to 675 MHz, respectively. The sensitivities and the operating frequency of the AlN/SiC-based sensors are lower than those shown in [30], but they are referred to a temperature compensated device, whose sensitivity and operating frequency can be increased by lowering the acoustic wavelength.

In reference [31] Wingqvist *et al.* demonstrate the remarkable gravimetric sensitivity (1,410 and 500 cm^2/g) of the SiO_2 and AlN sides of the temperature compensated S_0 mode sensor based on thin AlN(2 μm)/SiO_2(0.79 μm) membranes. The AlN/SiO_2-based sensor shows an operation frequency f_0 = 850 MHz and a K^2 as low as 0.8% for $\lambda = 10.5$ μm. These sensitivities are much higher than the highest value obtainable with the AlN/SiC-based sensors for the same wavelength ($\lambda = 10.5$ μm). It is important to note that, for equal λ, the thickness of the AlN/SiC plate configurations (from 3.5 to 7.1 μm) is much larger than that of the AlN/SiO_2 plate (2.79 μm) described in reference [31]. For AlN/SiC plates showing thicknesses comparable with that of the AlN/SiO_2 plate, (obtainable, for example, by using $\lambda = 4$ μm), a sensitivity as high as 910 cm^2/g is achievable on the SiC surface of the AlN(1.7 μm)/SiC(0.8 μm) sensor, for f_0 ~2 GHz and K^2 = 1.77%; a sensitivity as high as 961 cm^2/g is obtainable on the AlN surface of the AlN(1 μm)/SiC(0.4 μm) plate, for f_0 ~2.5 GHz and K^2 =2.17%. Hence, the temperature compensated AlN/SiC plates allow, for comparable layers thicknesses, lower sensitivity but higher operating frequency and K^2 than those obtainable with the temperature compensated AlN/SiO_2 plates. Thus, gravimetric sensors showing sensitivity higher than that obtainable with the SiC/AlN-based mass sensors do exist, but at the price of losing the temperature compensation and resistance to harsh environments. Hence the best choice for a practical system requires a compromise between some design parameters (such as the phase velocity, the thermal stability, the K^2, the S_m and R_m) that concur in the overall device performance evaluation.

The AlN/SiC-based devices can find other possible applications, such as energy harvesting, pressure and temperature sensing, especially at high temperatures and harsh environment, thanks to the AlN and SiC resistance to chemical attacks at elevated temperatures compared to their counterparts such as silicon and ZnO.

3. CONCLUSIONS

The propagation of the S_0 Lamb mode along 3C-SiC/c-AlN composite plates have been investigated by theoretical calculation with respect to the AlN and SiC films thickness, acoustic wave propagation direction, electrical boundary conditions and temperature. The phase velocity, the TCD and the K^2 dispersion curves of four AlN/SiC-based coupling configurations have been theoretically studied specifically addressing the design of temperature-compensated, enhanced-coupling, GHz-range electroacoustic devices. The sensitivity and limit of resolution of gravimetric sensors based on SiC/AlN plates have been theoretically studied with respect to the AlN and SiC sensing surface. The AlN/SiC-based mass sensors are proven to achieve remarkable performances (high sensitivity, low resolution values, thermostability and enhanced coupling efficiency) that are important prerequisite for the design of future devices, based on resonator principles, to be used in the context of chemical, biological and physical quantities detection.

REFERENCES

1. Levinshtein, M.E.; Rumyantsev, S.L.; Shur, M.S. *Properties of Advanced Semiconductor Materials: GaN, AIN, InN, BN, SiC, SiGe*; John Wiley & Sons, Inc.: New York, NY, USA, 2001.

2. Liu, F.; Carraro, C.; Chu, J.; Maboudian, R. Residual stress characterization of polycrystalline 3C-SiC films on Si(100) deposited, from methylsilane. *J. Appl. Phys.* **2009**, *106*.

3. Roper, C.S.; Howe, R.T.; Maboudian, R. Stress control of polycrystalline 3C-SiC films in a large-scale LPCVD reactor using 1,3-disilabutane and dichlorosilane as precursors. *J. Micromech. Microeng.* **2006**, *16*, 2736–2739.

4. Isshiki, T.; Nishio, K.; Abe, Y.; Komiyama, J.; Suzuki, S.; Nakanishi, H. Semipolar nitrides grown on Si(001) offcut substrates with 3C-SiC buffer layers. *Mater. Sci. Forum* **2009**, *600–603*, 1317–1320.

5. Tanaka, Y.; Hasebe, Y.; Inushima, T.; Sandhu, A.; Ohoya, S. Comparison of AlN thin films grown on sapphire and cubic-SiC substrates by LP-MOCVD. *J. Crystal Growth* **2000**, *209*, 410–414.

6. Lin, C.-M.; Lien, W.-C.; Felmetsger, V.V.; Hopcroft, M.A.; Senesky, D.G.; Pisano, A. AlN thin films grown on epitaxial 3C-SiC (100) for piezoelectric resonant devices. *App. Phys. Lett.* **2010**, *97*.

7. Lin, C.-M.; Chen, Y.-Y.; Pisano, A.P. Theoretical investigation of lamb wave characteristics in AlN/3C-SiC composite membranes. *Appl. Phys. Lett.* **2010**, *97*.

8. Adler, E.L.; Farnell, G.W.; Slaboszewicz, J.; Jen, C.K. Interactive PC Software for SAW Propagation in Anisotropic Multilayers. Proceedings of the IEEE Ultrasonics Symposium, San Diego, CA, USA, 27–29 October 1982; pp. 103–107.

9. Tsubuochi, K.; Sugai, K.; Mikoshiba, N. AlN Material Constants Evaluation and SAW Properties on AlN/Al_2O_3 and AlN/Si. Proceedings of the IEEE Ultrasonics Symposium, Chicago, IL, USA, 14–16 October 1981; pp. 375–380.

10. Hellwege, K.-H.; Hellwege, A.M. *Landolt-Börnstein, Numerical Data and Functional Relationships in Science and Technology, New Series, Group III*; Springer-Verlag: Berlin, Germany, 1979.

11. Farnell, G.W. Properties of elastic surface waves. In *Physical Acoustics, Principles and Methods*; Mason, W.P., Thurston, R.N., Eds.; Academic Press: New York, NY, USA, 1970; Volume VI, Chapter 4; p. 121.

12. Farnell, G.W.; Adler, E.L. Elastic wave propagation in thin layers. In *Physical Acoustics, Principles and Methods*; Mason, W.P., Thurston, R.N., Eds.; Academic Press: New York, NY, USA, 1972; Volume IX, Chapter 2; p. 79.

13. Cohen, M.G. Optical study of ultrasonic diffraction and focusing in anisotropic media. *J. Appl. Phys.* **1967**, *38*, 3821–3828.

14. Caliendo, C. Analysis of dispersive electroacoustic coupling configurations for application to gigahertz-band, temperature-compensated AlN-based acoustic devices. *Appl. Phys. Lett.* **2008**, *92*.

15. Li, Z.; Bradt, R.C. The single crystal elastic constants of hexagonal SiC to 1000 °C. *Int. J. High Technol. Ceram.* **1988**, *4*, 1–10.

16. Li, Z.; Bradt, R.C. The single-crystal elastic constants of cubic (3C) SiC to 1000° C. *J. Mater. Sci.* **1987**, *22*, 2557–2559.

17. Snead, L.L.; Nozawa, T.; Katoh, Y.; Byun, T.-S.; Kondo, S.; Petti, D.A. Handbook of SiC properties for fuel performance modeling. *J. Nucl. Mater.* **2007**, *371*, 329–377.

18. De Anna, R.G.; Roy, S.; Zorman, C.A.; Mehregany, M. Modeling of SiC Lateral Resonant Devices over a Broad Temperature Range. Proceedings of the International Conference on Modeling and Simulation of Microsystems, San Juan, Puerto Rico, 19–21 April 1999; p. 644.

19. Zhao, X.; Yang, F.; Zhang, H.; Xiao, P. Nondestructive evaluation of high-temperature elastic modulus of 3C-SiC using Raman scattering. *J. Raman Spectrosc.* **2012**, *43*, 945–948.

20. Hu, Y.F. Calculations for the structures and elastic properties of 3C-SiC compound at high temperature and pressure.*J. Sichuan Univ. Sci. Eng.* **2009**, *22*, 121–124.

21. Salem, J.; Li, Z.; Bradt, R.C. Thermal Expansion and Elastic Anisotropy in Single Crystal $A1_2O_3$ and SiC Reinforcements. NASA Technical Memorandum-TM-106516. Proceeding of ASME Symposium on Advances Composite Materials and Structures, American Society of Mechanical Engineers, Anaheim, CA, USA, 10–12 December 1986.

22. Caliendo, C. Analysis of the acoustoelectric behavior of microwave frequency, temperature-compensated AlN-based multilayer coupling configurations. *J. Appl. Phys.* **2008**, *104*.

23. Tabib-Azar, M. *Microactuators: Electrical, Magnetic, Thermal, Optical, Mechanical, Chemical and Smart Structures*; Kluwer Academic Press: Boston, MA, USA, 1998; pp. 139–140.

24. Choujaa, A.; Tirole, N.; Bonjour, C.; Martin, G.; Hauden, D.; Blind, P.; Cachard, A.; Pommier, C. AlN/silicon Lamb-wave microsensors for pressure and gravimetric measurements. *Sens. Actuators A Phys.* **1995**, *46–47*, 179–182.

25. Wang, Z.; Cheeke, J.D.N.; Jen, C.K. Perturbation method for analyzing mass sensitivity of planar multilayer acoustic sensors. *IEEE Trans. UFFC* **1996**, *43*, 844–851.

26. Auld, B.A. *Acoustic Fields and Waves in Solids*; John Wiley & Sons: New York, NY, USA, 1973; Volume 2.

27. Ballantine, D.S.; White, R.M.; Martin, S.J.; Ricco, A.J.; Zellers, E.T.; Fye, G.C.; Wohltjen, H. *Acoustic Wave Sensors: Theory, Design, & Physico-Chemical Applications*; Academic Press Limited: London, UK, 1997.

28. Caliendo, C. Theoretical investigation of high velocity, temperature compensated Rayleigh waves along AlN/SiC substrates for high sensitivity mass sensors. *Appl. Phys. Lett.* **2012**, *100*.

29. Lee, C.M.; Wong, K.M.; Chen, P.; Lau, K.M. GaN-Based Lamb-Wave Mass-Sensors on Silicon Substrates. Proceedings of the IEEE SENSORS 2010 Conference, Kona, HI, USA, 1–4 November 2010.

30. Arapan, L.; Alexieva, G.; Avramov, I.D.; Radeva, E.; Strashilov, V.; Yantchev, V.; Katardjiev, I. Highly mass-sensitive thin film plate acoustic resonators (FPAR). *Sensors* **2011**, *11*, 6942–6953.

31. Wingqvist, G.; Arapan, L.; Katardjiev, I. A micromachined thermally compensated thin film Lamb wave resonator for frequency control and sensing applications. *J. Micromech. Microeng.* 2009, *19*.

CHAPTER 5

On the Reliability of a Solitary Wave Based Transducer to Determine the Characteristics of Some Materials

Wen Deng [1,2,†], Amir Nasrollahi [2,†], Piervincenzo Rizzo [2,*,†] and Kaiyuan Li [2]

[1] School of Automation, Northwestern Polytechnical University, Xi'an 710072, China

[2] Laboratory for Nondestructive Evaluation and Structural Health Monitoring Studies, Department of Civil and Environmental Engineering, University of Pittsburgh, 3700 O'Hara Street, 729 Benedum Hall, Pittsburgh, PA 15261, USA

ABSTRACT

In the study presented in this article we investigated the feasibility and the reliability of a transducer design for the nondestructive evaluation (NDE) of the stiffness of structural materials. The NDE method is based on the propagation of highly nonlinear solitary waves (HNSWs) along a one-dimensional chain of spherical particles that is in contact with the material to be assessed. The chain is part of a built-in system designed and assembled to excite and detect HNSWs, and to exploit the dynamic interaction between the particles and the material to be inspected. This interaction influences the time-of-flight and the amplitude of the solitary pulses reflected at the transducer/material interface. The results of this study show that certain features of the waves are dependent on the modulus of elasticity of the material and that the built-in system is reliable. In the future the proposed NDE method may provide a cost-effective tool for the rapid assessment of materials' modulus.

KEYWORDS

highly nonlinear solitary waves; nondestructive evaluation; modulus of elasticity; concrete; magnetostrictive sensors

1. INTRODUCTION

In the last ten years interest in the use of highly nonlinear solitary waves (HNSWs) in physics and engineering has increased considerably. These waves are compact nondispersive mechanical waves that can form and travel in highly nonlinear systems, such as a closely packed chain of elastically interacting spherical particles [1,2,3] or in the tensegrity units [4,5]. Typically, solitary waves are induced by tapping the first particle of the chain with a particle striker identical to the spheres forming the chain.

In the study presented in this paper, we assembled four built-in systems, hereinafter indicated as the HNSW transducers, to excite and sense HNSWs. Each transducer consisted of a chain of spherical particles surmounted by an electromagnet to lift and release the striker, and a coil and two permanent magnets around the nine-th particle of the chain to sense the propagating wave using the magnetostriction principles. We then quantified the reliability of these transducers at inferring the modulus of a few materials placed in contact with the transducers. The method exploited the dynamic interaction between the solitary waves propagating within the chain and the test specimens.

Two sets of experiments were conducted. In the first one, the four transducers were tested against samples made of soft polyurethane, hard polyurethane, and steel. The cases of free transducers and transducers partially immersed in water were also considered to set a baseline data. The results were compared to assess the transducers repeatability, and were then compared to the numerical predictions obtained with a discrete particle model (DPM) to verify the accuracy of our instruments. The study on the dynamical interaction between solitary waves and bulk materials is not novel; Manciu and Sen [6] investigated the wave reflections from rigid wall boundaries. Falcon *et al.* [7] studied the collision of a column of N beads with a fixed wall. Job *et al.* [8] evaluated the collision of a single solitary wave with elastic walls with various stiffness. Yang *et al.* [9] demonstrated that the formation and propagation of reflected HNSWs depend on the modulus of elasticity and geometry of the adjacent medium. Finally, Ni *et al.* [10] monitored the curing of fresh cement using a HNSW-based transducer. Our group has also investigated the interaction of HNSWs with plates and slender beams [11,12,13,14]. In all the above studies, the sensing element was either a force sensor placed at one end of the chain or a sensor bead, *i.e.*, a piezo-element glued in between two half particles and inserted along the chain.

In the second investigation, the same approach was used to find the modulus of elasticity of hardened concrete. The findings were then compared to the results obtained from the conventional ultrasonic pulse velocity (UPV) method, and the destructive tests standardized in the ASTM C469, ASTM C39, and the ACI 318 [15,16,17]. The purpose of this second investigation was the development of a novel nondestructive evaluation (NDE) method to estimate the strength of concrete in existing structures.

Many NDE methods for concrete have been proposed, tested, and commercialized in the last three decades. The most common method is perhaps the one based on the measurement of the velocity of linear bulk ultrasonic waves propagating through concrete. Traditionally, commercial transducers are used to generate and detect ultrasonic bulk waves [18,19,20,21,22,23,24]. Parameters, such as wave speed and attenuation are theoretically correlated to the properties of the material. This approach is usually referred to as the UPV method, and it is extensively used in practice and in researches for monitoring the properties of hardened or fresh concrete. When the back wall of the sample is impractical, the wave reflection method is used, and impact-echo and pulse-echo methods are two common used techniques of this kind [25,26,27].

With respect to ultrasonic-based NDE, the proposed HNSW-based approach: (1) exploits the propagation of HNSWs confined within the grains; (2) employs a cost-effective transducer; (3) measures different parameters (time of flight, speed, and amplitude of one or two solitary pulses) that can be eventually used to correlate few concrete variables; (4) does not require any knowledge of the sample thickness; (5) does not require an access to the sample's back-wall. The solitary waves based method may resemble the Schmidt hammer that consists of a spring-driven steel hammer that hits the specimen with a defined energy. Part of the impact energy is absorbed by the plastic deformation of the specimen and transmitted to the specimen, and the remaining impact energy is rebounded. The rebound distance depends on the hardness of the specimen and the conditions of the surface. The harder is the surface, the shorter is the penetration time or depth; as a result, the higher is the rebound [28,29]. The differences between our method and the Schmidt hammer are: the hammer cannot be applied onto fresh concrete and cement; only one parameter, the rebound value, is used in the Schmidt hammer test, while multiple HNSWs features can, in principle, be exploited to assess the condition of the underlying material. The Schmidt hammer may induce plastic deformation or microcracks to the specimen, while the HNSW approach is purely nondestructive as there is not mechanical impact on the material under testing. The present paper is entirely novel with respect to paper [30]. There, one HNSW transducer was used to monitor the curing of one concrete cylinder. The transducer embedded two sensor beads within the chain of particles forming the transducer.

For all the above, the novelty of this paper is are: (1) more transducers were designed, assembled and tested to assess their repeatability; (2) the transducers carried a sensing element based on magnetostriction. This was reported in [30] for the first time in a study that compared three sensing elements: a conventional sensor bead; a magnetostrictive sensing approach; and a piezo-cylinder. However, no other studies considered magnetostriction as a possible tool to detect the solitary waves, and this paper investigates the reliability of magnetostrictive sensing for solitary waves; (3) a hardware system was designed and assembled to drive the transducers simultaneously without the need of manual plug-and-play; (4) hardened concrete was tested in order to investigate the feasibility of this novel NDE method to assess the strength of existing concrete.

2. BACKGROUND

This section describes the underlying basis of HNSWs, and it follows the symbolism adopted in [10,31], which in turn is based on the analytical formulation provided in several works including [1,3,31]. Figure 1 illustrates the general principles of the proposed method: a 1-D chain of spherical particles is in contact with the material to be assessed; the impact of a striker, *i.e.*, a particle of equal size and mass of the other particles composing the chain, generates a single pulse that propagates through the chain. The interaction between two adjacent beads is governed by the Hertz's law [32,33]:

$$F = A\delta^{3/2} \tag{1}$$

and it establishes a relationship between the compression force F of the granules and the closest approach of the particle centers. In Equation (1) the coefficient A is given by:

$$A = \begin{cases} A_c = E\sqrt{2a}/3(1-\nu^2) \\ A_w = \dfrac{4\sqrt{a}}{3}\left(\dfrac{1-\nu^2}{E} + \dfrac{1-\nu_w{}^2}{E_w}\right)^{-1} \end{cases} \tag{2}$$

where A_c refers to the case of particle-to-particle contact, and A_w refers to the case of particle-to-wall (infinite curvature radius) contact case. In Equation (2) E, a, and v are modulus of elasticity, radius, and Poisson's ratio of the spherical beads, respectively, whereas E_w and v_w are the modulus of elasticity and Poisson's ratio of the wall, respectively.

The combination of the nonlinear interaction (Equation (1)) and a zero tensile strength in the chain of spheres leads to the formation and propagation of compact solitary waves [33]. When the wavelength is much larger than the particles' diameter, the speed V_S of the waves depends on the maximum dynamic strain ξ_m [33] which, in turn, is related to the maximum forceF_m between the particles in the discrete chain [3]. When the chain of beads is under a static pre-compression force F_0 ($F_0 << F_m$), the initial strain of the system is referred to as ξ_0. It should be noted that in configurations like the one shown in Figure 1, the pre-compression is provided by the self-weight of the chain. The speed V_s has a nonlinear dependence on the normalized maximum strain $\xi_r =$

ξ_m/ξ_0, or on the normalized force $f_r = F_m/F_0$, expressed by the following equation [3]:

$$V_s = c_0 \frac{1}{(\xi_r - 1)} \times \left\{ \frac{4}{15} \left[3 + 2\xi_r^{5/2} - 5\xi_r \right] \right\}^{1/2}$$
$$= 0.9314 \left(\frac{4E^2 F_0}{a^2 \rho^3 \left(1 - v^2\right)^2} \right)^{1/6} \frac{1}{\left(f_r^{2/3} - 1\right)} \left\{ \frac{4}{15} \left[3 + 2f_r^{5/3} - 5f_r^{2/3} \right] \right\}^{1/2} \quad (3)$$

where c_0 is the wave speed in the chain initially compressed with a force F_0 in the limit $f_r = 1$, and ρ is the density of the material. When f_r (or ξ_r) is very large, Equation (3) becomes:

$$V_s = 0.6802 \left(\frac{2E}{a\rho^{3/2}\left(1 - v^2\right)} \right)^{1/3} F_m^{1/6} \quad (4)$$

and the shape of a solitary wave in a system moving with velocity V_s can be closely approximated as [33]:

$$\xi = \left(\frac{5V_s^2}{4c^2} \right) \cos^4 \left(\frac{\sqrt{10}}{5a} x \right) \quad (5)$$

where:

$$c = \sqrt{\frac{2E}{\pi\rho(1 - v^2)}} \quad (6)$$

and x is the local coordinate in the moving system.

The equations above predict that the mechanical properties of a specimen in contact with the chain of particles influence the contact stiffness of the chain-material interface. When a single incident solitary wave (ISW) reaches the interface with the material to be tested, the pulse is partially reflected giving rise to the primary reflected solitary wave (PSW). When the incident pulse interacts with a medium that is much softer than the particles carrying the acoustic energy, one or more secondary reflected solitary waves (SSWs) [7,9,10,29]. Many studies, including some from our group, determined that the characteristics of the reflected pulses such as amplitude and time of flight (TOF) are correlated to the modulus of the underlying material. Here, the TOF denotes the transit time at a given bead in the granular crystal between the incident and the reflected waves. The TOF of the PSW and SSW, hereinafter indicated as TOF_{PSW} and TOF_{SSW}, respectively, are typically estimated by measuring the arrival time of the peak amplitude of the ISW and the PSW or the SSW at a given particle.

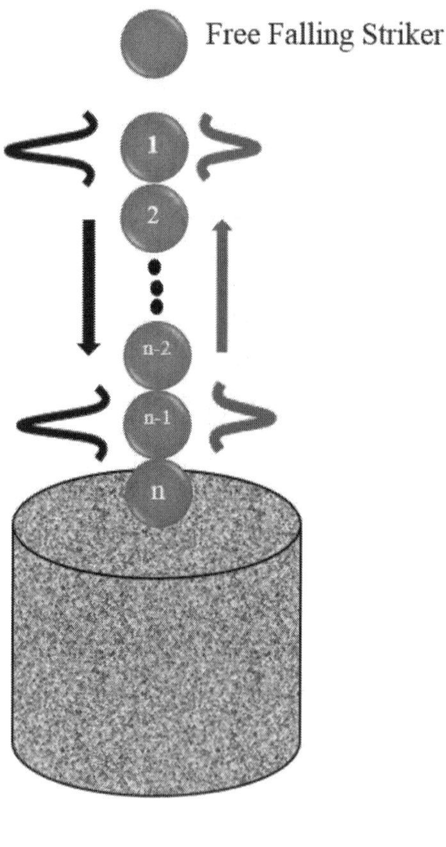

Figure 1. General scheme of the nondestructive evaluation method based on the propagation of highly nonlinear solitary waves.

3. EXPERIMENTAL SETUP

The spheres shown in Figure 1 were assembled into the transducer schematized in Figure 2. The chain consisted of 16 particles with $2a$ = 19.05 mm. The second bead from the top was made of nonferromagnetic steel (AISI 304, McMaster-Carr, Aurora, OH, USA) whereas the other particles were made of ferromagnetic low-carbon steel (AISI 1020, McMaster-Carr). The properties of these two steels are listed in Table 1. The chain was held by a Delrin Acetal Resin tube (8627K219, McMaster-Carr) with outer diameter D_0 = 22.30 mm and inner diameter slightly larger than $2a$ in order to minimize the friction between the striker and the inner wall of the tube and to minimize acoustic leakage from the chain to the tube. The striker was a ferromagnetic bead, and it was driven by an electromagnet. A constant axial magnetic field was created along the chain and centered at the 9th particle using two identical permanent bridge magnets (5841K55, McMaster-Carr). A coil was wrapped around the tube and around the magnetic field in order to attain a magnetostrictive sensor (MsS) utilized to measure the propagation of the solitary waves within the chain. Finally, a 0.254 mm thick aluminum sheet was glued to the tube's end to prevent the free fall of the particles.

Magnetostriction can be used to excite and detect stress waves using Faraday's law and the Villari effect, respectively. In this study, we used the Villari effect, which states that a pulse propagating through a ferromagnetic material modulates an existing magnetic field and generates voltage in the coils. In our experiments, the AISI 1020 particles were magnetostrictive material and they were subjected to the magnetic field induced by the magnets. One of the authors used MsS to excite and detected guided ultrasonic waves [34,35,36].

Four transducers were built and driven simultaneously by a National Instruments-PXI 1042Q (National Instruments Co., Austin, TX, USA) unit running in LabVIEW and a DC power supply (BK PRECISION 1672, B&K Precision Co., Yorba Linda, CA, USA). We used a Matrix Terminal Block (NI TB-2643) to branch the PXI output into four switch circuits. Figure 3a shows the diagram of the switch circuit. A Metal-Oxide-Semiconductor Field-Effect Transistor (MOSFET) was used for switching the electronic signals. In Figure 3, the symbols G, D, and S represent the Gate, Drain, and Source terminals of the MOSFET, respectively. Because the B (body) terminal of the MOSFET was connected to the source terminal, it is omitted in the diagram. EM1 to EM4 represents the four electromagnets mounted on the transducers 1 to 4, respectively. Switches 1 to 4 represent the digital switches. When one of them was set to 1, the switch was turned on. Figure 3b shows the NI-PXI utilized in the experiment; Figure 3c illustrates the DC Power used to provide the electromagnet with a direct voltage, and Figure 3d is the switch circuit with a MOSFET and two terminal blocks.

4. NUMERICAL SIMULATION OF HNSW-BASED TRANSDUCERS

A numerical model was implemented using the concepts presented in Section 2. The differential equation of motion of 1-D chain of beads can be determined using the Lagrangian description of particle dynamics:

$$m\partial_{tt}^2 u_n = A_n \left[u_{n-1} - u_n\right]_+^{3/2} - A_n \left[u_n - u_{n+1}\right]_+^{3/2} \tag{7}$$

where u_i is the ith particle displacement, $[x]_+$ denotes $max(x,0)$, and A_n is what was defined in Equation (2). It was assumed that the displacement of wall in contact with the grains was zero. For the sake of brevity, the full details are not given here, but the interested reader is referred to paper [30]. With respect to work [30] our model did not include dissipation.

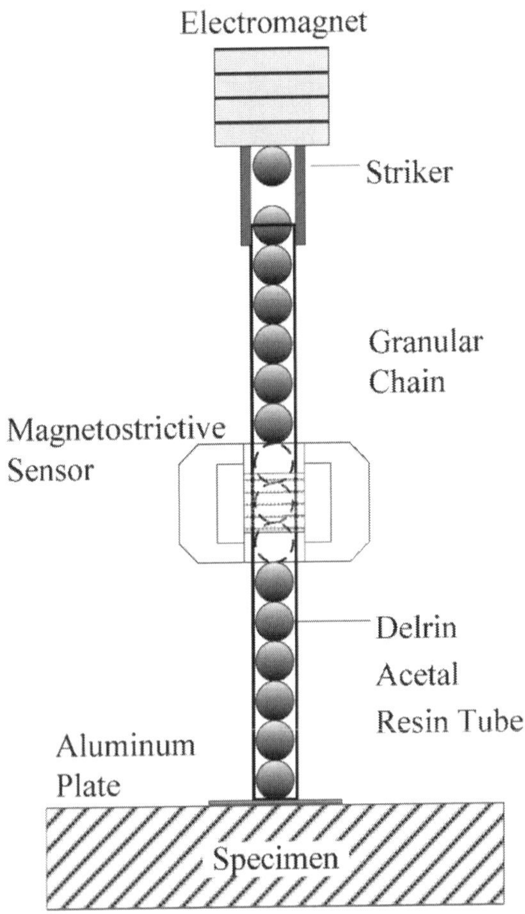

Figure 2. Schematic of the HNSWs-based transducer.

By solving Equation (7), the time-history of the particles' oscillation was obtained. Once the time-history of the displacement is determined from this equation, the force profile can be determined using Equation (1). To enable the comparison between the numerical results and the experimental findings, the model mirrored the experimental setup, and the material in contact with the last particle of the chain was considered to be the same as the experiments. The mechanical properties of AISI 304 steel [37], AISI 1020 steel [38], hard polyurethane [39], and soft polyurethane [39] are listed in Table 1. The effect of the aluminum sheet at the bottom of the tubes was added to the numerical simulation based on the theory of membranes [40,41]. The presence of the magnetic force generated by the two permanent magnets was included by considering three particles per coil subjected to a static compressive force equal to 1.0 N. This value was estimated by comparing the experimental wave speed of the incident solitary wave to the analytical prediction. For the case of the free transducers, the interface was the aluminum sheet in contact with the last particle. For the case of the transducers partially immersed in water, the contact effect was the sum of the sheet's stiffness and the pressure provided by a 5 mm water column, which was the same as in the experiments. Because in the experiments the coil of the MsS was wrapped around the 9th particle, all the results of the numerical simulation presented here are based on the wave features measured at the center of the same particle.

Figure 4 shows the numerical force profiles when the five different interfaces were considered. To ease the readability of the graph, the profiles were offset arbitrarily by 150 N. The top two plots represent the baseline profiles. The first pulse is the ISW. The second pulse is the PSW and it shows two small humps; this phenomenon occurs when the interface material or element is too weak to prevent the separation of the last few particles from the rest of the chain. When they bounce back to the original rest position, they collide with the chain at slightly different times giving rise to the two humps. Overall, Figure 4 demonstrates that the amplitude and the speed of both reflected waves are dependent on the material which is in contact with the chain: as the material's stiffness increases, the amplitude of the SSW decreases while the amplitude of the PSW increases, and their TOF_{PSW} and TOF_{SSW} diminish.

Table 1. Mechanical properties of the materials used as the beads and interface.

Material	Density (kg/m^3)	Modulus of Elasticity (GPa)	Poisson's ratio
Stainless steel (AISI 304) [37]	8000	200	0.29
Stainless steel (AISI 1020) [38]	7860	200	0.29
Soft polyurethane (SAWBONES, #1522-03) [39]	320	0.210	0.30
Hard polyurethane (SAWBONES, #1522-05) [39]	640	0.759	0.30

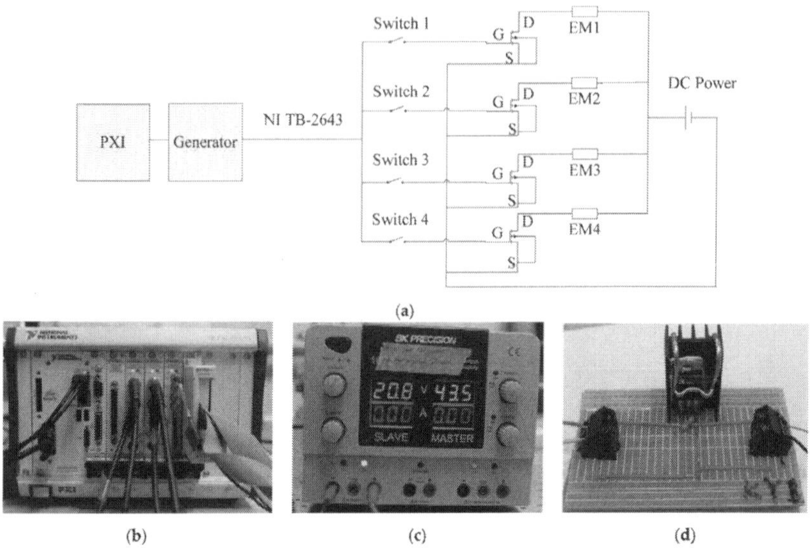

Figure 3. Experimental Setup. (**a**) The switch circuit; (**b**) The NI-PXI used to drive the transducers, and digitalize, and store the time waveforms; (**c**) The DC Power supply used to activate the electromagnets; (**d**) The switch circuit with a MOSFET and two terminal blocks.

To generalize the findings of Figure 4, Figure 5a displays the TOF_{PSW} as a function of the dynamic modulus of elasticity and the Poisson ratio of the contact material. It is observed that the Poisson's ratio has a small effect on the travel time whereas the modulus of elasticity has a significant impact on the time of flight of the PSW when $E_w < 100$ GPa. Figure 5b shows the ratio of the amplitude of the PSW to the amplitude of the ISW, hereinafter *PSW/ISW*, as a function of the modulus of elasticity and the Poisson ratio. Similar to the parameter of the TOF_{PSW}, the Poisson ratio does not have a significant impact on the ratio. Moreover, *PSW/ISW* steadily increases with respect to the modulus of elasticity until the modulus is about 50 GPa, and then it remains constant. Similar to Figure 5b, Figure 5c shows the *SSW/ISW* as a function of the materials' mechanical properties. The surface is complementary to the surface seen in Figure 5b. This implies that when the stiffness of the contact material is low, a part of the energy carried by the incident pulse spent to generate the SSW. Interestingly, above 50 GPa, the secondary solitary wave is not triggered, and based on the plot of Figure 5b, more than 90% of the incident acoustic energy is reflected back to the chain in the form of PSW.

Figure 6 shows the effect of the Poisson ratio on the TOF at three given values of the modulus of elasticity. The figure demonstrates that the effect of the Poisson ratio on the predicted TOF is extremely small. For example, at any given E, the variation of the TOF when the Poisson ratio varies from 0.1 to 0.3 is about 0.14%.

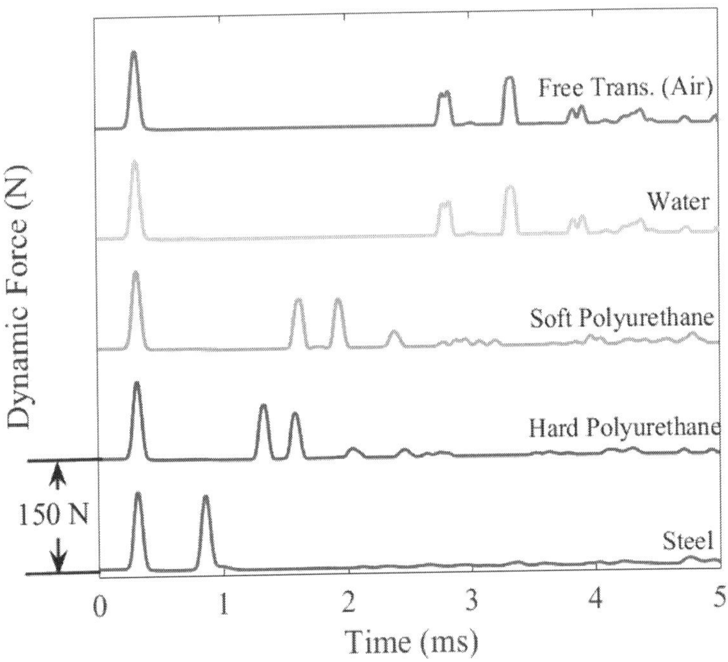

Figure 4. Numerical Results. Force profile measured at the center of the 9th bead.

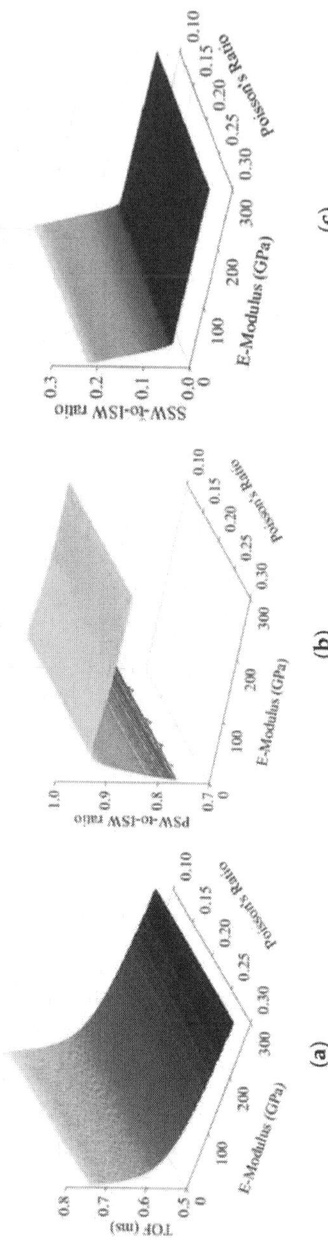

Figure 5. Numerical Results. (**a**) Time of flight of the primary solitary reflected wave; (**b**) PSW-to-ISW amplitude ratio; (**c**) SSW-to-ISW amplitude ratio as a function of the modulus of elasticity and Poisson's ratio of the material in contact with the chain.

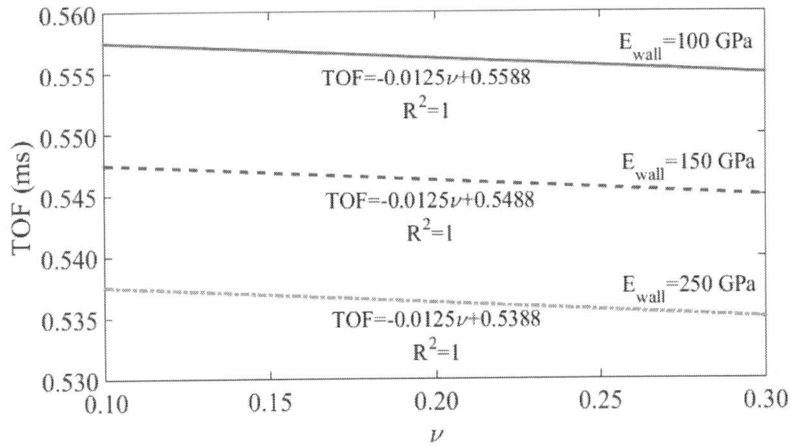

Figure 6. TOF as a function of Poisson's ratio for some values of E_{wall}.

5. EXPERIMENTAL RESULTS

Figure 7 shows one of the four transducers suspended in air, *i.e.*, the free transducer, partially immersed, and in contact with the soft polyurethane, hard polyurethane, and stainless steel block.

To assess the repeatability of the method and to verify that all four transducers were identical, one hundred measurements were taken for each medium for every transducer; hence, a total of 2000 waveforms were analyzed. Figure 8a–e shows the average of the 100 measurements for the five different media. To ease the readability of the panels, the waveforms were arbitrarily offset by 0.6 V.

The time waveforms show both positive and negative voltage. When the solitary wave travels through the constant magnetic field induced by the permanent magnets, it increases the compression between two adjacent particles, and it creates a positive gradient of the magnetic flux, which in turn induces the positive voltage.

When the pulse moves away, the dynamic compression disappears, a negative gradient of the magnetic flux is induced, and therefore, the output voltage has a negative gradient. In the figures, the first, second, and third pulses correspond to the ISW, PSW, and SSW, respectively.

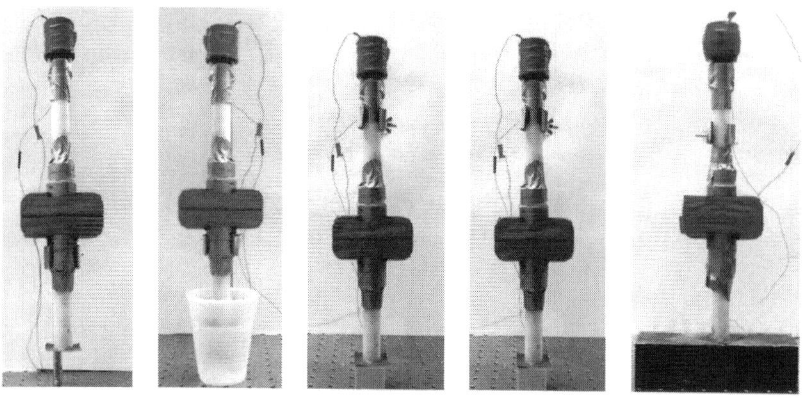

Figure 7. Photos of one HNSW-based transducer. From left to right: free transducer; partially immersed transducer; in contact with a soft polyurethane cube; in contact with a hard polyurethane cube; in contact with a stainless steel block.

The integral of the voltage *V*(*t*), measured at the center of the 9th particle, is proportional to the dynamic contact force *F*,*i.e.*, [30]:

$$F = K_c \int V(t)dt \tag{8}$$

where K_c is a conversion factor expressed in $N/V \cdot \mu s$.

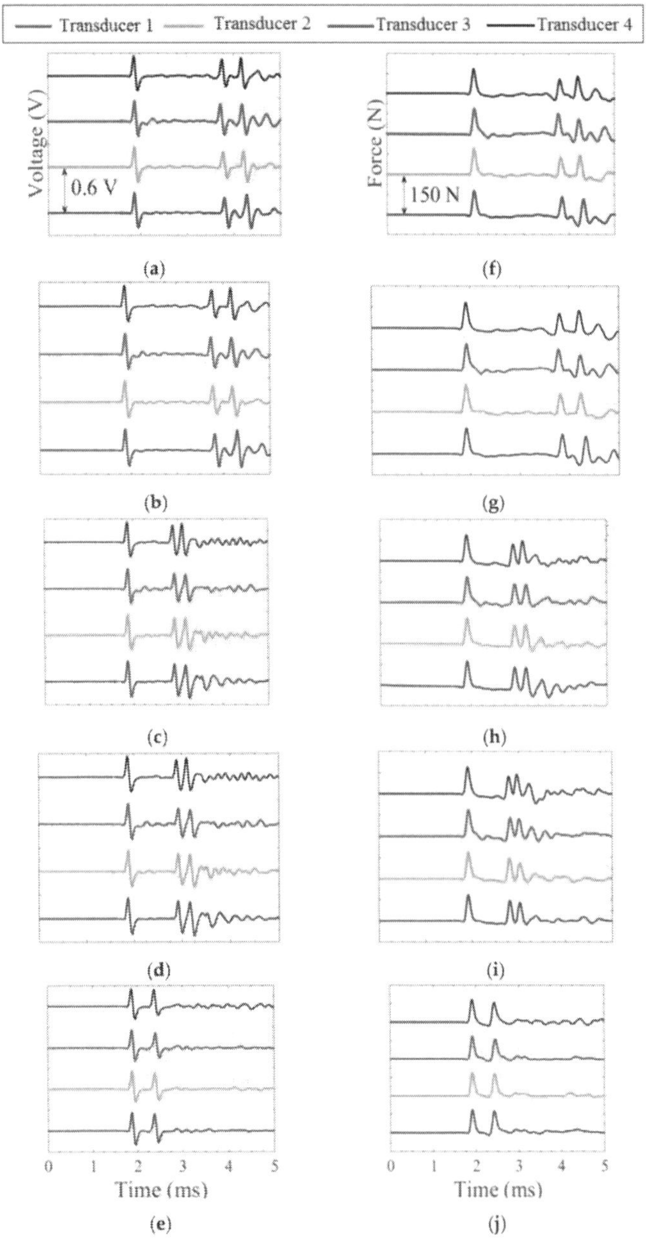

Figure 8. Experimental results. Average time waveforms (**left column**) and corresponding force profile (**right column**) measured by the four transducers under the five test scenarios. (**a**,**f**): air, (**b**,**g**): water, (**c**,**h**): soft polyurethane, (**d**,**i**): hard polyurethane, (**e**,**j**): stainless steel.

Under the assumption that the numerical model presented in Section 4 mirrored the experimental setup, we calculated $K_c$by computing the ratio of the numerical dynamic force to the experimental time integral both associated with the ISW, which is the only pulse not affected by the presence of any underlying material. For each transducer, the 500 experimental pulses were considered. The results are listed in Table 2, and they reveal that the values of K_c are very similar. In fact, the difference between the smallest (6.42 N/V·μs) and the largest (6.67 N/V·μs) value is below 4% which proves the consistency of the assembled transducers. Moreover, within each transducer, the corresponding coefficients of variation, *i.e.*, the ratio of the standard deviation σ to the mean value, are all below 5%, which is a good indicator of the repeatability of the incident pulse.

The values of K_c were then used in Equation (8) to infer the dynamic force profile from the experimental time waveforms.Figure 8f–j represents the experimental force for each transducer in contact with the five different media. The ISW, PSW, and the SSW are visible. The latter is absent when steel was tested. The graphs confirm the repeatability of the transducers and the excellent agreement with the numerical predictions presented in Figure 4.

To quantify the experimental data, Figure 9 shows the values related to the TOF_{PSW} and TOF_{SSW} measured by the four transducers for the five interfaces. The dots represent the average of the 100 measurements whereas the vertical bars represent the 2σ. As expected from the analytical prediction, the TOF_{PSW} and TOF_{SSW} decrease with the increase of the stiffness of the interfaces. The four transducers provide almost identical results; the largest difference was below 6.0% when the steel was tested, and the smallest difference was equal to 0.0011% occurred for the free transducers. Moreover, the small standard deviations confirm the high repeatability of the measurements.

Table 2. Conversion factors K_c to compute the dynamic contact force.

Transducer Number	Conversion Factor (N/V· μs)	Standard Deviation (N/V· μs)	Coefficient of Variation (%)
1	6.58	0.1664	2.53
2	6.42	0.1689	4.10
3	6.67	0.0947	1.42
4	6.67	0.1334	2.00

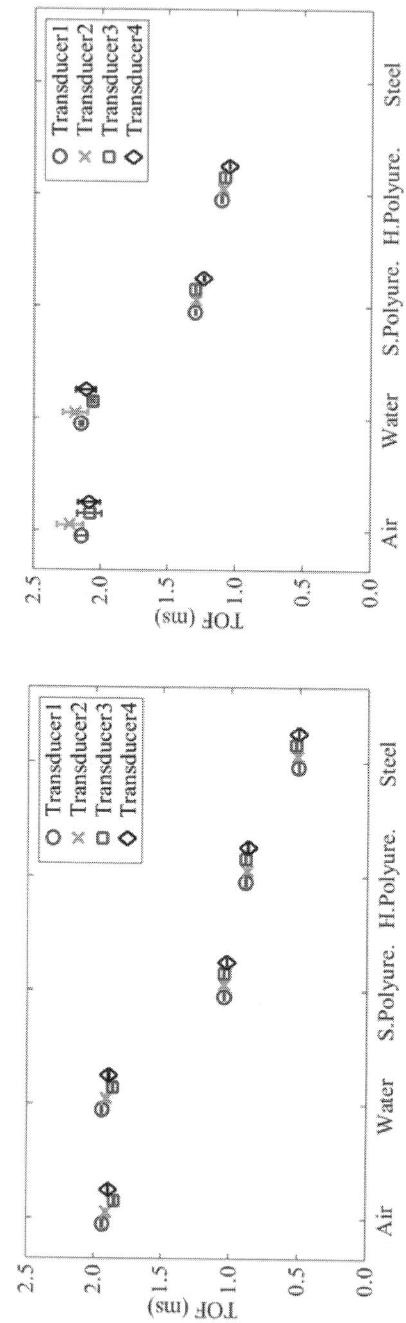

Figure 9. Time of flight of both (**a**) PSW and (**b**) SSW and the corresponding standard deviation as a function of the five different media.

Table 3 compares the experimental and the numerical TOF_{PSW}. The differences between the numerical simulation and the experiments range between 0.43% and 18.6%. The largest discrepancies are relative to the soft and the hard polyurethane. These differences stem mainly from viscoelastic behavior [9] of the polyurethane in the chain which was not modeled in the numerical simulation.

6. CONCRETE SLABS TESTS

6.1. Material Characteristics

After proving the reliability of the transducers, we tested eight 30 × 30 × 5 cm^3 concrete slabs in order to study the capability of the novel technology to determine the modulus of elasticity of six-month-old hardened concrete. The concrete mix was designed with w/c = 0.42, and fine and coarse river gravel aggregates. The average mass of the slabs was 10.93 ± 0.21 kg. Three cylinders made of the same batch were cast, cured for 28 days, and tested according to standard test methods to estimate the static modulus of elasticity and the Poisson ratio of concrete. These values, namely $E_s^{ASTM\ C469}$ and ν, are listed in Table 4.

Section 8.5.1 of ACI 318 provides an empirical relationship between the modulus of elasticity E_s^{ACI} and the ultimate compressive strength f'_c. Typical values for concrete span from f'_c = 13.8 MPa (2 ksi) to 48.3 MPa (7 ksi) [42]. Since we obtained the compressive strength of the concrete mix according to ASTM C39, we can estimate the modulus of elasticityE_s^{ACI} as:

$$E_s^{ACI} = 33(\rho^{1.5})\sqrt{f_c'} \quad (9)$$

$$E_s^{ACI} = 0.0427(\rho^{1.5})\sqrt{f_c'} \text{ (in SI units)} \quad (10)$$

Table 3. Numerical and experimental time of flight relative to the primary reflected wave when the chain is in contact with the five media.

Medium	Numerical TOF_{PSW} (ms)	Experimental TOF_{PSW} (ms)	Num. and Exp. Difference (%)
Air	2.1350	1.9791	7.88
Water	2.1350	1.9837	7.62
Soft Polyurethane	1.3275	1.0802	18.63
Hard Polyurethane	1.0250	0.9237	9.88
Steel	0.5525	0.5549	0.43

Table 4. Values of the compressive strength, the modulus of elasticity, and the Poisson ratio obtained from the ASTM C39, ASTM C469 Tests, and ACI 318 correlation.

Samples	$E_s^{ASTMC469}$ (GPa)	ν	f'_c (MPa)	E_s^{ACI} (GPa)
1	28.9	0.158	43.3	33.7
2	25.5	0.164	41.0	32.8
3	26.5	0.155	41.5	33.0

In Equation (10), ρ is the density of concrete in kg/m^3, and f'_c is the compressive strength in MPa. The density of concrete is measured, and the compressive strength f'_c was obtained from the ASTM C39 at 28 days. The result is included in Table 4.

6.2. HNSW-Based Method: Setup and Results

Figure 10 shows the experimental setup. The four HNSW-based transducers were placed above each specimen, and each specimen was tested at four different locations simultaneously, in order to average local surface heterogeneities. For each transducer, 100 measurements were taken using the same hardware/software system described in Section 3. It is noted again that the tests were carried 6 months after casting the samples.

Figure 11 shows the force profile obtained after averaging the 100 measurements on slab 4. The presence of the ISW and PSW is clearly visible. The small hump tracing the PSW is the SSW. Despite the inhomogeneous nature of concrete and the use of different transducers, the waveforms are almost identical.

Figure 10. Concrete test setup. Photo of the four transducers placed above one of the eight samples.

Figure 12 displays the average TOF_{PSW} and TOF_{SSW}, and the corresponding standard deviation. The scatter of the features' values is below 5%, and it is likely due to the intrinsic slight differences among the transducers and the heterogeneity of the samples' surface. The empirical values of the TOF_{PSW} and the average Poisson ratio obtained with the ASTM C469 (ν = 0.16) were used to derive the empirical dynamic modulus of elasticity (E_d^{HNSW}) from Figure 5a. It is remarked here that the selection of the Poisson ratio could have been done using typical concrete values without affecting the results significantly. Table 5 reports the E_d^{HNSW} computed from all four transducers for all eight samples. By looking at the eight readings from the same transducer, it is evident that the scatter of the data is much higher than the variability of the individual transducer. This suggests that the heterogeneous surface of the samples might have scattered the data. To support this hypothesis, Figure 13 displays a close-up view of the surface of slab 1, slab 3, and slab 4. Many voids are visible on slabs 1 and 3. These voids (bugholes) typically result from the migration of entrapped air and to a less extent water, to the fresh concrete interfaces [43,44]. These voids and the roughness of the surfaces may weaken the local stiffness of the concrete or may influence the dynamic interaction at the material/transducer interface.

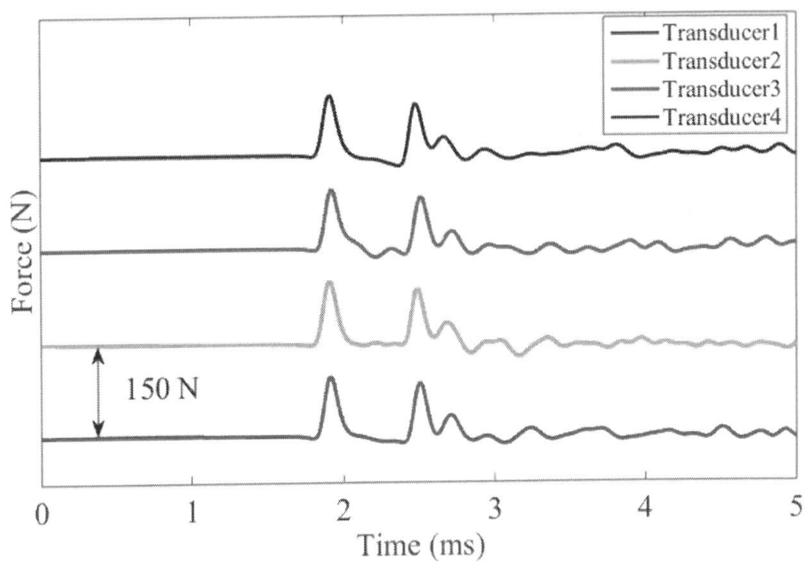

Figure 11. Experimental force profile measured by the four transducers as a function of time when testing one of the eight slabs.

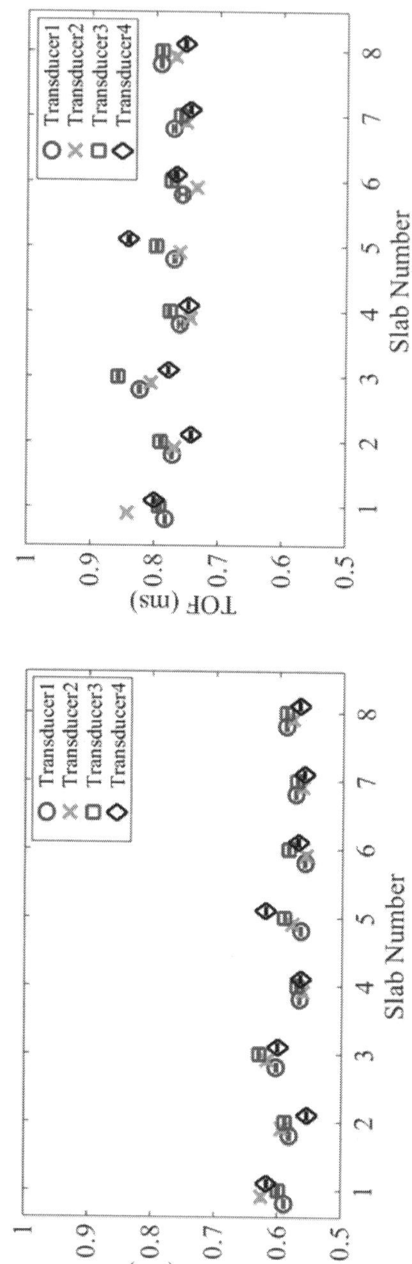

Figure 12. Average time of flight and the relative standard deviations of both (**a**) PSW and (**b**) SSW obtained from the four transducers for the eight concrete slabs.

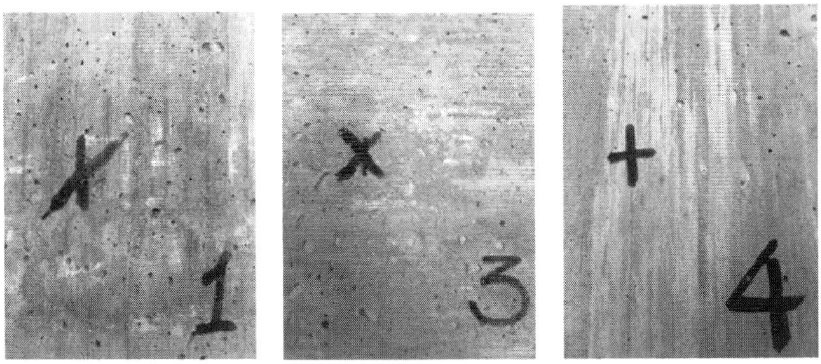

(a) (b) (c)

Figure 13. Close-up view of the surface of (**a**) slab 1, (**b**) slab 3, and (**c**) slab 4.

The dynamic modulus E_d^{HNSW} was converted into the static modulus of elasticity (E_s^{HNSW}) using the empirical formula proposed by Lydon and Balendran [45,46]:

$$E_s^{HNSW} = 0.83E_d^{HNSW} \tag{11}$$

The results are also presented in Table 5. The superscript emphasizes the fact that these values were obtained with the novel NDE method. With the exception of the outliers measured from slabs 1 and 3, the average E_d^{HNSW} is equal to 30.8 GPa and its corresponding static modulus is equal to 25.6 GPa.

6.3. Ultrasonic Pulse Velocity (UPV) Method: Setup and Results

For comparative purposes we applied the UPV method using the setup presented in Figure 14. A function generator was used to excite a 5-cycle, 2 V peak-to-peak, 500 kHz sine wave. Two V103-RM transducers (Olympus, Tokyo, Japan) were used to transmit and receive the waves, respectively. Two Olympus 5660C amplifiers were used to amplify the signals. To mimic the setup with the solitary waves, we probed five different locations as shown in Figure 14b. The received signals were digitized with an oscilloscope. Figure 14c shows an example of the transmitted and the detected waveforms.

It is known that for elastic homogeneous solid media, the longitudinal wave velocity is given by:

$$V = \sqrt{\frac{KE_d^{UPV}}{\rho}} \quad (12)$$

where $K = (1 - v)/((1 + v)\cdot(1 - 2v))$, and ρ is density of the medium. The static value of Poisson's ratio obtained from destructive test can be utilized in the absence of the dynamic Poisson's ratio [26]. To adhere to the results associated with the solitary wave testing, we considered $v = 0.16$. From Equation (12):

$$E_d^{UPV} = V^2\rho/K \quad (13)$$

The ultrasonic wave velocity was calculated by considering the arrival time difference between the first peaks of the transmitted and the received signals. The pulse velocity, the dynamic modulus of elasticity computed from Equation (13), and the static modulus derived from Equation (11) are listed in Table 6. Each value is the average of the five measurements taken from each slab. The results show that the variation among the moduli is very small, and the slabs 1 and 3 are not outliers. The reason is that the UPV test measures the pulse velocity of the waves traveling through the specimen, and therefore, it is less susceptible to surface conditions.

Table 5. Concrete test. The TOF_{PSW}, the predicted dynamic modulus of elasticity (E_d), and the corresponding static modulus of elasticity (E_s) of the slabs measured by the HNSW-based transducers.

Slab Number	Sensor Number	TOF_{PSW} (ms)	Modulus of Elasticity (GPa)		
			E_d^{HNSW}	Average E_d^{HNSW}	Average E_s^{HNSW}
1	1	0.6250	22.9	17.4*	14.4
	2	0.6477	16.8		
	3	0.6475	16.8		
	4	0.6650	13.1		
2	1	0.6225	26.1	28.5	23.7
	2	0.6193	28.8		
	3	0.6350	20.3		
	4	0.6001	39.0		
3	1	0.6400	18.8	16.3*	13.5
	2	0.6442	18.7		
	3	0.6786	11.1		
	4	0.6486	16.8		
4	1	0.6075	35.0	37.8	31.4
	2	0.5827	59.8		
	3	0.6211	26.2		
	4	0.6125	30.2		
5	1	0.6050	35.0	26.5	22.0
	2	0.6000	39.0		
	3	0.6386	19.5		
	4	0.6685	12.6		
6	1	0.5975	41.1	38.4	31.9
	2	0.5800	64.0		
	3	0.6336	21.2		
	4	0.6186	27.5		
7	1	0.6125	30.1	28.7	23.8
	2	0.6175	27.5		
	3	0.6250	23.9		
	4	0.6075	33.3		
8	1	0.6275	23.9	24.9	20.7
	2	0.6175	27.4		
	3	0.6375	19.5		
	4	0.6150	28.8		
Average		0.6250	27.3	30.8*	25.6*

* The results of Slabs 1 and 3 were not used in the average.

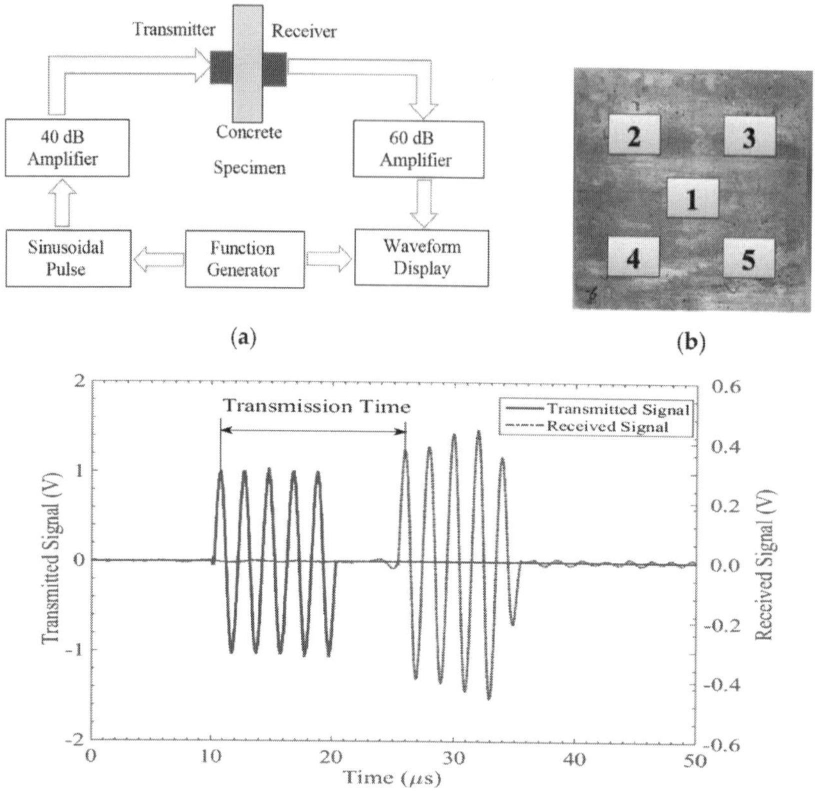

(c)

Figure 14. Experimental setup of the UPV test. (**a**) Overall schematic; (**b**) The tested areas on a concrete slab. (**c**) Typical time waveform of the transmitted (blue line) and the received (red dash line) signals.

7. DISCUSSION AND CONCLUSIONS

This article shows the working principles and the reliability of a transducer able to generate and detect HNSWs and the capability of a novel NDE method based on the properties of these waves at assessing the mechanical properties of the materials to be inspected.

Table 6. UPV test. Velocity of the longitudinal bulk wave and predicted dynamic and static modulus of elasticity.

Slab Number	Average Bulk Wave Velocity (m/s)	E_d^{UPV} (GPa)	E_s^{UPV} (GPa)
1	3386	24.9	20.7
2	3646	29.1	24.2
3	3531	26.9	22.3
4	3387	24	19.9
5	3530	26.4	21.9
6	3436	25.3	21
7	3524	26.4	21.9
8	3637	27.7	23
Average	3510	26.3	21.8

The moduli obtained from the different testing methods are presented in Table 7 and displayed in Figure 15. The HNSWs-based prediction ranges from 13.5 GPa to 31.9 GPa, the value for modulus from the UPV test range from 19.9 GPa to 24.2 GPa, and the modulus calculated from the empirical equation range from 32.8 GPa to 33.7 GPa.

Table 7. Concrete tests. Estimated static modulus of elasticity of the test samples using two destructive and two nondestructive methods.

Slab Number	Nondestructive Methods	
	EsHNSW (GPa)	EsUPV (GPa)
1	14.4*	20.7
2	23.7	24.2
3	13.5*	22.3
4	31.4	19.9
5	22.0	21.9
6	31.9	21.0
7	23.8	21.9
8	20.7	23.0
Average	25.6	21.8
Sample	Destructive Methods	
	EsASTM C469 (GPa)	EsACI (GPa)
1	28.9	33.7
2	25.5	32.8
3	26.5	33.0
Average	26.9	33.1

* The results of slabs 1 and 3 were not used in the average.

In Figure 15, the average values and the corresponding standard deviations are presented. There are small differences in the modulus of elasticity of the slabs measured by the UPV test. The differences among the eight slabs are mainly caused by the following three reasons: first, the rough surface of the specimen changes the thicknesses or the path lengths in different testing areas, and this problem results in inaccurate path length ΔL when calculated the pulse velocity; second, the aggregates are important factor for the pulse velocity. Their sizes, types, and content can affect the test results; third, the rough surface results in insufficient contact with the commercial transducers, and a weaker pulse velocity will be received by the transducer; consequently, the onset of received signals may be difficult to be figured out. Based on the UPV measurements, the average dynamic modulus of elasticity of the eight concrete slabs is 26.3 GPa, and its corresponding static modulus is equal to 21.8 GPa.

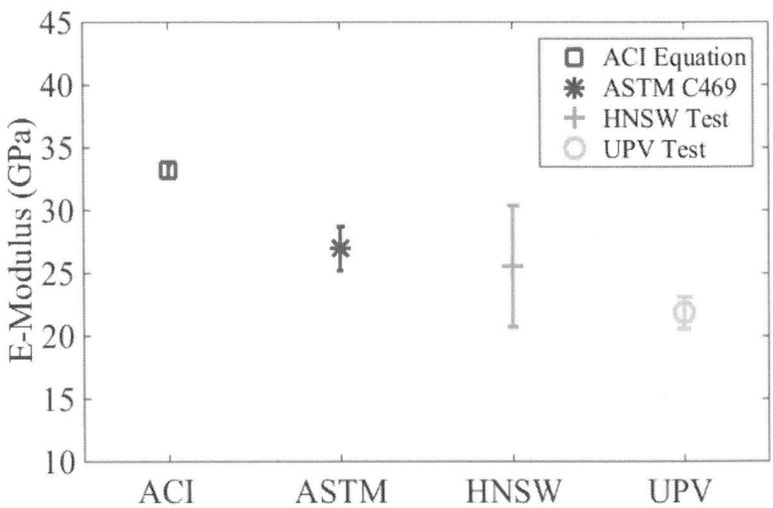

Figure 15. The average estimated static modulus of elasticity and the corresponding standard deviations of the four different methods.

The results relative to the ASTM C469 conducted after 28 days from casting, the modulus of elasticity of the batch ranges from 25.5 to 28.9 GPa. Figure 15 shows that the standard deviation of the empirical equation is the smallest among the four methods.

Furthermore, the modulus of elasticity estimated via the UPV is smaller than the other methods, while HNSW-based transducers provide the static modulus of elasticity very close to the ASTM C469. The difference between HNSW-based method and the ASTM is about 5%, while the difference between ASTM and the UPV method and the empirical equation of the ACI 318 are 20% and 23%, respectively.

The main advantages of operating HNSW-based test is that its implementation is fast and easy; hence, it is possible to carry out a large number of tests and take the average of the value of modulus of elasticity estimated from each observation. Another advantage of the HNSW-based method for evaluating the modulus of elasticity of the specimens is that the value obtained by this method is less susceptible to any damage and/or the presence of reinforcing steels existing inside the element. For example, cracks and dense reinforcing steel may cause refraction and dispersion of the ultrasonic waves while this problem does not affect the results of the HNSW-based transducers. Future studies shall include more testing to verify the effect of concrete mix design on the properties of the solitary waves. An effort shall also be made to mitigate the effect of the concrete surface at the contact point with the chain.

ACKNOWLEDGMENTS

The study presented in this paper is supported by partial technical results drawn from the currently active work order PITT WO 008, funded by the Pennsylvania Department of Transportation and the Federal Highway Administration under contract 4400011482. The first author conducted this research as a visiting scholar at the University of Pittsburgh under a sponsorship from the China Scholarship Council. Support for the fourth author was provided by the U.S. National Science Foundation, Grant CMMI 1200259.

We acknowledge Andrew Bunger and his research group at the Department of Civil and Environmental Engineering at the University of Pittsburgh for lending the concrete samples and for providing the results of the standard tests.

The use of the results or reliance on the material presented is the responsibility of the reader. The contents of this document are not meant to represent standards and are not intended for use as a reference in specifications, contracts, regulations, statutes, or any other legal document. The opinions and interpretations expressed are those of the authors and other duly referenced sources. The views and findings reported herein are solely those of the writers and not necessarily those of the US Department of Transportation, Federal Highway Administration, or the Commonwealth of Pennsylvania. This paper does not constitute a standard, a specification, or regulations.

AUTHOR CONTRIBUTIONS

P. Rizzo conceived and designed the experiments; Wen Deng and Kaiyuan Li performed the experiments; A. Nasrollahi analyzed the data and implemented the numerical model; W. Deng, A. Nasrollahi, and P. Rizzo wrote the paper.

CONFLICTS OF INTEREST

The authors declare no conflict of interest.

REFERENCES

1. Nesterenko, V.F. Propagation of nonlinear compression pulses in granular media. *J. Appl. Mech. Tech. Phys.* **1983**, *24*, 733–743.

2. Coste, C.; Falcon, E.; Fauve, S. Solitary waves in a chain of beads under Hertz contact. *Phys. Rev. E* **1997**, *56*.

3. Daraio, C.; Nesterenko, V.F.; Herbold, E.; Jin, S. Tunability of solitary wave properties in one-dimensional strongly nonlinear phononic crystals. *Phys. Rev. E* **2006**, *73*.

4. Fraternali, F.; Senatore, L.; Daraio, C. Solitary waves on tensegrity lattices. *J. Mech. Phys. Solids* **2012**, *60*, 1137–1144.

5. Fraternali, F.; Carpentieri, G.; Amendola, A.; Skelton, R.E.; Nesterenko, V.F. Multiscale tunability of solitary wave dynamics in tensegrity metamaterials. *Appl. Phys. Lett.* **2014**, *105*.

6. Manciu, F.S.; Sen, S. Secondary solitary wave formation in systems with generalized Hertz interactions. *Phys. Rev. E***2002**, *66*.

7. Falcon, E.; Laroche, C.; Fauve, S.; Coste, C. Collision of a 1-D column of beads with a wall. *Eur. Phys. J. B Cond. Matter Complex Syst.* **1998**, *5*, 111–131.

8. Job, S.; Melo, F.; Sokolow, A.; Sen, S. How Hertzian solitary waves interact with boundaries in a 1D granular medium.*Phys. Rev. Lett.* **2005**, *94*.

9. Yang, J.; Silvestro, C.; Khatri, D.; De Nardo, L.; Daraio, C. Interaction of highly nonlinear solitary waves with linear elastic media. *Phys. Rev. E* **2011**, *83*.

10. Ni, X.; Rizzo, P.; Yang, J.; Katri, D.; Daraio, C. Monitoring the hydration of cement using highly nonlinear solitary waves. *NDT&E Int.* **2012**, *52*, 76–85.

11. Cai, L.; Rizzo, P.; Al-Nazer, L. On the coupling mechanism between nonlinear solitary waves and slender beams. *Int. J. Solids Struct.* **2013**, *50*, 4173–4183.

12. Bagheri, A.; La Malfa Ribolla, E.; Rizzo, P.; Al-Nazer, L.; Giambanco, G. On the use of l-shaped granular chains for the assessment of thermal stress in slender structures. *Exp. Mech.* **2014**, *55*, 543–558.

13. Ni, X.; Rizzo, P. Use of highly nonlinear solitary waves in NDT. *Mat. Eval.* **2012**, *70*, 561–569.

14. Ni, X.; Rizzo, P. Highly nonlinear solitary waves for the inspection of adhesive joints. *Exp. Mech.* **2012**, *52*, 1493–1501.

15. ASTM C469—Standard Test Method for Static Modulus of Elasticity and Poisson's Ratio of Concrete in Compression. Avalible online: http://www.astm.org/Standards/C469.htm (accessed on 21 December 2015).

16. ASTM C39—Standard Test Method for Compressive Strength of Cylindrical Concrete Specimens. Avalible online: http://www.astm.org/Standards/C39.htm (accessed on 21 December 2015).

17. ACI Committee. *Building Code Requirements for Structural Concrete (ACI 318-08) and Commentary (ACI 318R-08)*; American Concrete Institute: Cambridge, MA, USA, 2008.

18. Karaiskos, G.; Deraemaeker, A.; Aggelis, D.; Van Hemelrijck, D. Monitoring of concrete structures using the ultrasonic pulse velocity method. *Smart Mater. Struct.* **2015**, *24*.

19. Shah, A.; Ribakov, Y.; Zhang, C. Efficiency and sensitivity of linear and non-linear ultrasonics to identifying micro and macro-scale defects in concrete. *Mater. Design* **2013**, *50*, 905–916.

20. Ye, G.; Lura, P.; Van Breugel, K.; Fraaij, A. Study on the development of the microstructure in cement-based materials by means of numerical simulation and ultrasonic pulse velocity measurement. *Cement Concrete Comp.* **2004**, *26*, 491–497.

21. Lin, Y.; Lai, C.-P.; Yen, T. Prediction of ultrasonic pulse velocity (UPV) in concrete. *ACI Mater. J.* **2003**, *100*, 21–28.

22. Amini, K.; Lesak, A.; Sohal, A.; Adato, J.; Delatte, N. In using ultrasonic pulse velocity to predict properties and performance of pervious concrete. In Proceedings of the Transportation Research Board 93rd Annual Meeting, Washingron DC, WA, USA, 12–16 January 2014.

23. Shariq, M.; Prasad, J.; Masood, A. Studies in ultrasonic pulse velocity of concrete containing GGBFS. *Constr. Build. Mater.* **2013**, *40*, 944–950.

24. Popovics, S.; Rose, J.L.; Popovics, J.S. The behaviour of ultrasonic pulses in concrete. *Cement Concrete Res.* **1990**, *20*, 259–270.

25. McCann, D.; Forde, M. Review of NDT methods in the assessment of concrete and masonry structures. *NDT&E Int.***2001**, *34*, 71–84.

26. Malhotra, V.M.; Carino, N.J. *Handbook on Nondestructive Testing of Concrete*, 2nd ed; CRC Press: Boca Raton, FL, USA, 2003.

27. Tinkey, Y.; Olson, L.; Wiggenhauser, H. Impact echo scanning for discontinuity detection and imaging in posttensioned

concrete bridges and other structures. *Mater. Eval.* **2005**, *63*, 64–69.

28. Rapoport, J.R.; Popovics, J.S.; Kolluru, S.V.; Shah, S.P. Using ultrasound to monitor stiffening process of concrete with admixtures. *ACI Mater. J.* **2000**, *97*, 675–685.

29. Aydin, A.; Basu, A. The Schmidt hammer in rock material characterization. *Eng. Geol.* **2005**, *81*, 1–14.

30. Ni, X.; Cai, L.; Rizzo, P. A comparative study on three different transducers for the measurement of nonlinear solitary waves. *Sensors* **2013**, *13*, 1231–1246.

31. Ni, X.; Rizzo, P.; Daraio, C. Laser-based excitation of nonlinear solitary waves in a chain of particles. *Phys. Rev. E* **2011**,*84*.

32. Hertz, H. On the contact of elastic solids. *J. Reine Angew. Math.* **1881**, *92*, 156–171.

33. Nesterenko, V.F. *Dynamics of Heterogeneous Materials*; Springer Science & Business Media: New York, NY, USA, 2013.

34. Lanza di Scalea, F.; Rizzo, P.; Seible, F. Stress measurement and defect detection in steel strands by guided stress waves. *J. Mater. Civ. Eng.* **2003**, *15*, 219–227.

35. Rizzo, P.; Lanza di Scalea, F. Feature extraction for defect detection in strands by guided ultrasonic waves. *Struct. Health Monit.* **2006**, *5*, 297–308.

36. Rizzo, P.; Lanza di Scalea, F. Wavelet-based feature extraction for automatic defect classification in strands by ultrasonic structural monitoring. *Smart Struct. Syst.* **2006**, *2*, 253–274.

37. AISI Type 304 Stainless Steel. Available online: http://asm.matweb.com/search/SpecificMaterial.asp?bassnum=MQ304A (accessed on 21 December 2015).

38. AISI Type 1020 Stainless Steel. Available online: http://www.azom.com/article.aspx?ArticleID=9145#3 (accessed on 21 December 2015).

39. SAWBONES. Biomechanical Test Materials. Available online: http://www.sawbones.com/UserFiles/Docs/biomechanical_catalog.pdf. (accessed on 21 December 2015).

40. Timoshenko, S.; Woinowsky-Krieger, S. *Theory of Plates and Shells*; McGraw-hill New York: New York, NY, USA, 1959; Volume 2.

41. Eaton, W.P.; Bitsie, F.; Smith, J.H.; Plummer, D.W. In a new analytical solution for diaphragm deflection and its application to a surface micromachined pressure sensor. In Proceedings of the International Conference on Modeling and Simulation of Microsystems, San Juan, Puerto Rico, 19–21 April 1999.

42. Voigt, T.; Shah, S.P. Properties of early-age portland cement mortar monitored with shear wave reflection method.*ACI Mater. J.* **2004**, *101*, 473–482.

43. Bugholes. Available online: http://www.cement.org/for-concrete-books-learning/concrete-technology/concrete-construction/bugholes (accessed on 21 December 2015).

44. Ozkul, T.; Kucuk, I. Design and optimization of an instrument for measuring bughole rating of concrete surfaces. *J. Franklin Inst.* **2011**, *348*, 1377–1392.

45. Lydon, F.; Balendran, R. Some observations on elastic properties of plain concrete. *Cement Concrete Res.* **1986**, *16*, 314–324.

46. Popovics, J.; Zemajtis, J.; Shkolnik, I. *A Study of Static and Dynamic Modulus of Elasticity of Concrete, ACI-CRC Final Report*; Civil and Environmental Engineering, University of Illinois: Urbana, IL, USA, 2008.

CHAPTER 6

Stress Wave Propagation in Viscoelastic-Plastic Rock-Like Materials

Liu Lang [1,2], KI-IL Song [3,*], Yue Zhai [4], Dezheng Lao [5] and Hang-Lo Lee [3]

[1] Energy School, Xi'an University of Science and Technology, Xi'an 710054, China

[2] Key Laboratory of Western Mines and Hazards Prevention, Ministry of Education of China, Xi'an 710054, China

[3] Department of Civil Engineering, Inha University, Incheon 402-751, Korea

[4] School of Geology Engineering and Geomatics, Changan University, Xi'an 710054, China

[5] School of Civil & Resource Engineering, University of Western Australia, Perth 6009, Australia

ABSTRACT

Rock-like materials are composites that can be regarded as a mixture composed of elastic, plastic, and viscous components. They exhibit viscoelastic-plastic behavior under a high-strain-rate loading according to element model theory. This paper presents an analytical solution for stress wave propagation in viscoelastic-plastic rock-like materials under a high-strain-rate loading and verifies the solution through an experimental test. A constitutive equation of viscoelastic-plastic rock-like materials was first established, and then kinematic and kinetic equations were then solved to derive the analytic solution for stress wave propagation in viscoelastic-plastic rock-like materials. An experimental test using the SHPB (Split Hopkinson Pressure Bar) for a concrete specimen was conducted to obtain a stress-strain curve under a high-strain-rate loading. Inverse analysis based on differential evolution was conducted to estimate undetermined variables for constitutive equations. Finally, the relationship between the attenuation factor and the strain rate in viscoelastic-plastic rock-like materials was investigated. According to the results, the frequency of the stress wave, viscosity coefficient, modulus of elasticity, and density play dominant roles in the attenuation of the stress wave. The attenuation decreases with increasing strain rate, demonstrating strongly strain-dependent attenuation in viscoelastic-plastic rock-like materials.

KEYWORDS

viscoelastic-plastic rock-like materials; constitutive equation; stress wave equation; strain rate; attenuation; SHPB apparatus

1. INTRODUCTION

Previous studies have examined the propagation of the stress wave in composite materials, such as rock and concrete under a loading for the seismic design of various infrastructure systems [1,2,3].

The propagation of the stress wave is governed mainly by the inherent physical properties of the material and characteristics of the stress wave. Although brittle rock-like materials (e.g., fresh granite) are typically heterogeneous and show brittle fracture behaviors under a static load and low-strain-rate load, their dynamic behaviors are dependent on the strain rate under a high-strain-rate load [4,5,6]. In general, mechanical responses of the materials dependent on the strain rate can be divided into two categories: time-independent transient responses (responses not dependent on the strain rate) and time-related non-transient responses (responses dependent on the strain rate) [7]. From a macroscopic perspective, the dispersion phenomenon is equivalent to the absorption phenomenon in that the two phenomena are commonly induced by the dissipative force of viscous materials from the viscosity effects.

An analytical solution for stress wave propagation in viscoelastic materials is used widely in polymer materials based on the viscosity effects according to the elastic responses [8]. An analytical solution for stress wave propagation in viscoplastic materials has been developed by considering the viscosity effects based on the plastic responses [9]. A theoretical solution for stress wave propagation in viscoelastic-plastic materials that considers the viscosity effects on both elastic and plastic parts has been developed based on the elastic-plastic responses [10].

2. ANALYTIC SOLUTION FOR STRESS WAVE PROPAGATION IN VISCOELASTIC-PLASTIC ROCK-LIKE MATERIALS

2.1. Constitutive Equation of Viscoelastic-Plastic Rock-Like Materials

To derive a constitutive equation for viscoelastic-plastic rock-like materials under a high-strain-rate load, such a material should be assumed to be a composite material composed of elastic, plastic, and viscous components. In addition, these three components should be assumed to be distributed randomly and that the mixture reflects the macroscopic dynamic and mechanical behaviors of viscoelastic-plastic rock-like materials. According to element model theory [11,12,13], which is widely accepted in the constitutive equation of rock-like materials, the constitutive model of viscoelastic-plastic rock-like materials can be established.

This model is composed of three main components: an elastomer, a Newtonian body, and a St. Venant body. Figure 1shows the constitutive model of viscoelastic-plastic rock-like materials, and the constitutive equations are presented as follows:

(1) Elastomer. The stress–strain relationship is linear elastic. The constitutive equation is:

$$\sigma_E = E_1 \varepsilon_E \tag{1}$$

where E_1 is the initial elastic modulus of the elastic component, ε_E and σ_E are the strain and stress of the elastomer, respectively.

(2) Newtonian body. The stress is proportional to the strain; there is no immediate deformation, and the constitutive equation can be defined as:

$$\sigma_N = \eta \dot{\varepsilon}_N \quad (2)$$

where η is the viscosity coefficient, $\dot{\varepsilon}_N$ and σ_N are the strain rate and stress of the Newtonian body, respectively.

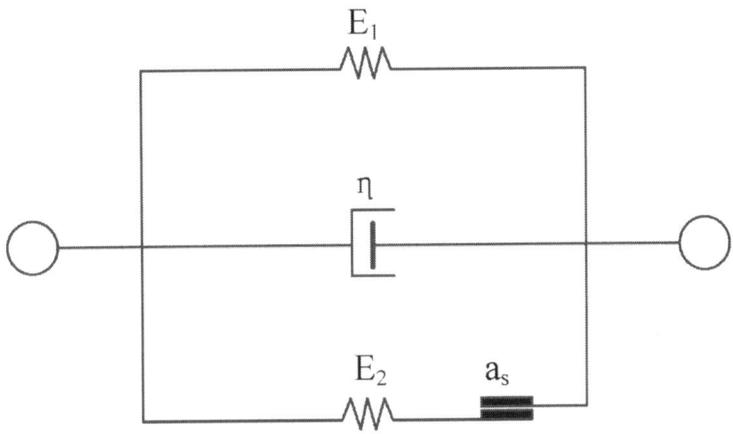

Figure 1. Configuration of the constitutive model for viscoelastic-plastic rock-like materials. Note: a_s is St. Venant element.

(3) St. Venant body. To be ideal elastic-plastic, there should be no creep, relaxation, and springback. The constitutive equation is:

$$\sigma_{SV} = E_2 \varepsilon_{SV}, \ \sigma_{SV} < \sigma_S \quad (3a)$$

$$\varepsilon_S \to \infty, \ \sigma_{SV} \geqslant \sigma_S \quad (3b)$$

where E_2 is the initial elastic modulus of the St. Venant body and σ_S is the elastic limit. The elastic limit is the lowest stress level at which the material begins to deform plastically. The elastic limit should be determined experimentally. If the stress on the specimen exceeds the elastic limit, the stress-strain curve becomes nonlinear because of the plastic strain component [2].

According to element model theory and the parallel relationship between three stress components, the stress and strain should satisfy the following:

$$\sigma = \sigma_E + \sigma_{SV} + \sigma_N \tag{4a}$$

$$\varepsilon = \varepsilon_E = \varepsilon_{SV} = \varepsilon_N \tag{4b}$$

From the three sets of constitutive equations of the model elements and their parallel relationship, differential equations of the constitutive model for viscoelastic-plastic rock-like materials can be derived as:

$$\sigma = (E_0 + E_2)\,\varepsilon + \eta\left(\frac{E_0 + E_1 + E_2}{E_1}\right)\dot{\varepsilon} - \frac{\eta\dot{\sigma}}{E_1},\ \sigma < \sigma_S \tag{5a}$$

$$\sigma = E_0\varepsilon + \eta\left(\frac{E_0 + E_1}{E_1}\right)\dot{\varepsilon} - \frac{\eta\dot{\sigma}}{E_1} + E_2\varepsilon_S,\ \sigma \geqslant \sigma_S \tag{5b}$$

Equation (5) is a special case of the constitutive equations for viscoelastic-plastic rock-like materials and can be rewritten in a general form as follows:

$$\alpha_0\sigma + \alpha_0\dot{\sigma} = b_0\varepsilon + b_{10}\dot{\varepsilon} \tag{6}$$

This equation is called a standard linear solid model. To describe the general behaviors of viscoelastic-plastic rock-like materials, a generalized linear viscoelastic-plastic model should be defined as:

$$\sum_{i=0}^{m} a_i \frac{\partial^i \sigma}{\partial t^i} = \sum_{i=0}^{n} b_i \frac{\partial^i \varepsilon}{\partial t^i} \tag{7}$$

where a_i and b_i are the material parameters.

2.2. The Wave Equation of Viscoelastic-Plastic Rock-Like Materials

This study confines stress wave propagation along the slender bar. Stress wave propagation along the slender bar can be simulated experimentally using an SHPB device. Through this test, the dynamic properties of a specimen material can be obtained [14,15,16,17]. The fundamental theory of the stress wave propagation in a slender bar can also be used to test the pile integrity.

To establish a stress wave equation for viscoelastic-plastic materials, kinematical and kinetic equations are incorporated with a constitutive equation of viscoelastic-plastic rock-like materials [18]. In the coordinate system, the central axis of a slender rock bar is taken as the *x*-axis, and the length is set to far exceed the diameter of the bar. Basically, A0 (the initial cross-sectional area prior to deformation), ρ0 (initial density), and the material parameters have no relationships with the coordinate system. In the remainder of this paper, *u* (displacement of the material point), ε (strain), ν (particle velocity), and σ (stress) are the longitudinal components along the *x*-axis.

The kinematic equation is often called a continuity equation or a mass conservation equation. A compatibility equation about ε and *u* can be established based on the monotropic continuity condition of displacement *u*:

$$\frac{\partial u}{\partial X} = \frac{\partial \varepsilon}{\partial t} \tag{8}$$

The kinetic equation is called an equation of motion or a momentum conservation equation. The following equation can be established based on the continuity condition as follows:

$$\rho_0 \frac{\partial v}{\partial t} = \frac{\partial \sigma}{\partial X} \tag{9}$$

A constitutive equation (physical equation) presents a continuous function between strain and stress. A piecewise continuous function can also be used in a constitutive equation in a manner similar to that in Equation (5).

Because the velocity of the stress wave induced by impact loading in hard rock materials is high and the travel time in the microelement is too short to cause heat exchange with the neighboring microelement, the process of impact loading can be considered to be approximately adiabatic. Therefore, it is unnecessary to implement an energy conservation equation to derive a closed-form solution for σ, ε, and *v*.

If viscoelastic-plastic rock-like materials are subject to a high-strain-rate compressive loading, then the stress wave is fully reflected in the fractured surface formed by the damage induced by the shock wave. Therefore, the stress wave cannot propagate forward. To ensure the stress function σ(ε) as a continuous differentiable function, this paper focuses mainly on the mechanical properties of viscoelastic-plastic rock-like materials without cracks. Under such a condition, the stress does not exceed the yield stress. Therefore, $Ed = d\sigma/d\varepsilon \geq 0$, and the stress wave velocity can be defined as follows:

$$C = \pm\sqrt{\frac{1}{\rho_0}\frac{d\sigma}{d\varepsilon}} = \sqrt{\frac{1}{\rho_0}E_d} \tag{10}$$

where *C* is the stress wave front velocity. The sine stress is a monotropic function of the strain; *C* is a monotropic function of strain; and ρ0C is the impedance of the material, which plays a decisive role in stress wave propagation. In Equation (10), the positive sign corresponds to the wave traveling from right to left, and *vice versa* for the negative sign.

The stress wave equation for viscoelastic-plastic rock-like materials can be obtained by solving kinematic and kinetic equations (Equations (8) and (9), respectively) and the constitutive equation (Equation (5)) for viscoelastic-plastic rock-like materials. If u (displacement) is considered to be an unknown quantity, then the calculation process is as follows:

If $\sigma<\sigma s$, then substitute $\varepsilon = \frac{\partial u}{\partial X}$ and $v = \frac{\partial u}{\partial t}$ into Equation (5a) separately and take the first derivative with respect to the x-axis on both the left and right sides:

$$\frac{\partial \sigma}{\partial X} = (E_0 + E_2)\frac{\partial^2 u}{\partial X^2} + \eta\left(\frac{E_0 + E_1 + E_2}{E_1}\right)\frac{\partial^3 u}{\partial t \partial X^2} - \frac{\eta \partial^2 \sigma}{E_1 \partial t \partial X} \tag{11}$$

Substituting Equation (7) into Equation (9) results in the following equation:

$$\rho_0 \frac{\partial^2 u}{\partial t^2} = (E_1 + E_2)\frac{\partial^2 u}{\partial X^2} + \eta \frac{\partial^3 u}{\partial t \partial X^{2'}} \quad \sigma < \sigma_s \tag{12}$$

This equation is a third-order partial differential equation of u. If the travel time of the stress wave is short, then the viscosity term containing the third derivative can be ignored. Therefore, the above equation can be expressed as a second-order partial differential equation for stress wave propagation along the elastic bar.

Similarly,

$$\rho_0 \frac{\partial^2 u}{\partial t^2} = E_1 \frac{\partial^2 u}{\partial X^2} + \eta \frac{\partial^3 u}{\partial t \partial X^{2'}} \quad \sigma \geqslant \sigma_s \tag{13}$$

Because of the viscosity term, the elastic wave equation u=F(X+C0t) does not satisfy the stress wave equation (Equation (14)). To solve this problem, a harmonic wave solution can be adopted:

$$u\left(X,t\right)=A\cdot\exp\left[i\left(\omega t-k_iX\right)\right] \tag{14}$$

Substituting Equation (14) into Equation (12) yields:

$$\rho_0\omega^2-\left(E_1+E_2\right)k_i^2-\eta i\omega k_i^2=0 \tag{15}$$

Because this is an equation with a complex number, the frequency ω can be considered a real number, and ki must be a complex number:

$$k_i=k+i\alpha \tag{16}$$

Substituting Equation (16) into Equation (15) yields:

$$\rho_0\omega^2-\left(E_1+E_2\right)\left(k^2-\alpha^2\right)-2ki\alpha\left(E_1+E_2\right)-\eta i\omega\left(k^2-\alpha^2\right)+2\eta\alpha k\omega=0 \tag{17}$$

If the real component is set to be equal to the imaginary component, then the following equations can be derived:

$$\rho_0\omega^2-\left(E_1+E_2\right)\left(k^2-\alpha^2\right)+2\eta\alpha k\omega=0 \tag{18}$$

$$\left(E_1+E_2\right)2k\alpha+\eta\omega\left(k^2-\alpha^2\right)=0 \tag{19}$$

Solving Equations (18) and (19) with respect to the attenuation factor α and the wave number *k* gives the following:

$$\alpha^2=\frac{\rho_0\omega^2\sqrt{\left(E_1+E_2\right)^2+\eta^2\omega^2}-\left(E_1+E_2\right)}{2[\eta^2\omega^2+\left(E_1+E_2\right)^2]} \tag{20a}$$

$$k^2 = \frac{\rho_0 \omega^2 \sqrt{(E_1 + E_2)^2 + \eta^2 \omega^2} + (E_1 + E_2)}{2[\eta^2 \omega^2 + (E_1 + E_2)^2]} \tag{20b}$$

The attenuation factor α can be either positive or negative. Because a positive root means that the amplitude increases with increasing propagation distance, it violates the law of energy conservation and has no physical meaning. Therefore, the attenuation factor α should be negative. The solution to Equation (12) can be obtained as follows:

$$u(X,t) = A\exp(-\alpha X)\exp[i(\omega t - kX)] \tag{21}$$

Similarly, if $\sigma \geq \sigma s$, then the harmonic wave solution in Equation (14) is the solution to Equation (13), and its form is similar to Equation (20). Under this condition, the wave number *k* and the attenuation factor α can be determined, respectively, using the following equations:

$$\alpha^2 = \frac{\rho_0 \omega^2 \left(\sqrt{E_1^2 + \eta^2 \omega^2} - E_1\right)}{2\left(E_1^2 + \eta^2 \omega^2\right)} \tag{22a}$$

$$k^2 = \frac{\rho_0 \omega^2 \left(\sqrt{E_1^2 + \eta^2 \omega^2} + E_1\right)}{2\left(E_1^2 + \eta^2 \omega^2\right)} \tag{22b}$$

Equations (20) and (22) can be rewritten as:

$$\alpha^2 = \frac{\rho_0 \omega^2 \left(\sqrt{b^2 + 1} - b\right)}{2\eta\left(b^2 + 1\right)} \tag{23a}$$

$$k^2 = \frac{\rho_0 \omega^2 \left(\sqrt{b^2 + 1} + b\right)}{2\eta\left(b^2 + 1\right)} \tag{23b}$$

where $b = \frac{E}{\eta\omega}$, E is the nominal elastic modulus. Here, E=E1+E2 if σ<σs, and E=E1 if σ≥σs, respectively.

Therefore, the elastic modulus of the material decreases when the stress exceeds the plastic threshold. The stress wave is influenced significantly by the attenuation factor under such a condition.

The particle velocity and strain of materials can be defined using spatial and temporal positions:

$$v(X,t) = \frac{\partial u(X,t)}{\partial t} = A\exp(-\alpha X)\exp[i(\omega t - kX)]\, i\omega = i\omega u(X,t) \tag{24}$$

$$\varepsilon(X,t) = \frac{\partial u(X,t)}{\partial t} = -i(\alpha + k)A\exp(-\alpha X)\exp[i(\omega t - kX)] = -i(\alpha + k)u(X,t) \tag{25}$$

The attenuation speed depends on the attenuation factor of the stress wave α, and its unit is [dB/m]. From Equations (20a) and (22a), the analytical solution for stress wave propagation in viscoelastic-plastic rock-like materials suggests that the attenuation factor α can be determined by the viscous coefficient of the material, the elastic modulus, density, and the frequency of the stress wave.

3. VERIFICATION OF THE ANALYTIC SOLUTION FOR STRESS WAVE PROPAGATION

3.1. An Experimental Test Using the SHPB and an Inverse Analysis

The mean strain rate $\dot{\varepsilon}$ and the elastic limit σs in the constitutive equation, Equation (5) can be obtained experimentally. The elastic modulus (E_1 and E_2) and the viscosity coefficient η are parameters to be determined. To obtain these unknown parameters, an experimental test using the SHPB (Split Hopkinson Pressure Bar) was adopted. The SHPB is an experimental technology to examine the dynamic mechanical properties of engineering materials. The pressure rod is assumed to be a simple elastic rod. Owing to impact load at one end, the elastic stress waves are propagated and cause deformation (*i.e.*, elastic deformation and plastic deformation).

This study verified the analytic solution derived in the previous section by conducting an SHPB test with concrete samples composed of fine aggregates. The height and diameter of the sample were 25 mm and 50 mm, respectively. The density of the sample was 2080 $kg{\cdot}m^{-3}$, and the maximum grain size was 2.5 mm. Figure 2 presents the SHPB apparatus. Based on the SHPB test of viscoelastic-plastic materials, the effects of the viscous coefficient, the elastic modulus, density, and the frequency of the stress wave on the attenuation factor were investigated. The stress-strain curves for the concrete specimen for three strain rates are presented in Figure 3.

The elastic stress wave along the guide rod in the SHPB can be decomposed into a series of harmonic components using the spectrum analysis method. The frequency and wavelength of harmonics are different from each other, and the harmonics propagate at different phase velocities [18]. Because the wavelength of the stress wave depends on the bullet length, the frequency of the stress wave varies, even for the same strain rate [19]. Dynamic properties of materials were not yet verified to be related directly to the frequency of the stress wave. Therefore, the effect of the frequency was not considered in the present study for parameter identification purposes.

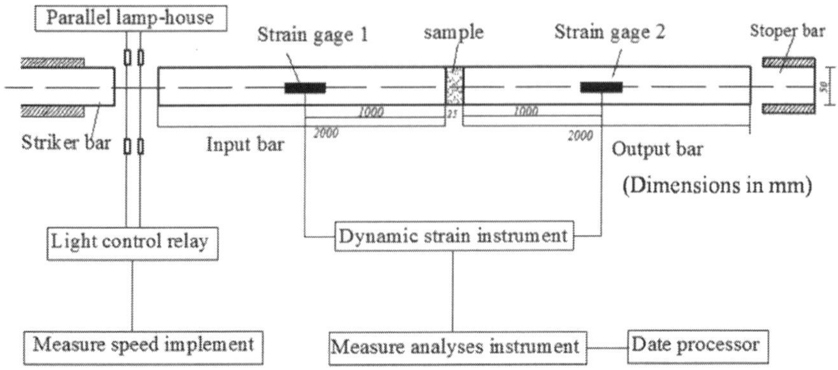

Figure 2. SHPB apparatus.

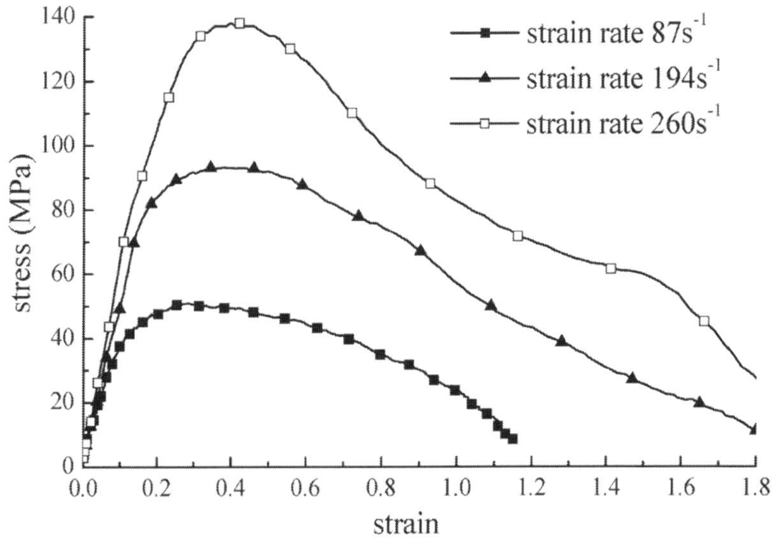

Figure 3. Stress-strain curve depending on the strain rate.

The identification process of unknown parameters using experimental test results is an optimization problem. An optimal solution providing the minimum mean square error in the objective function can be obtained as:

$$\hat{\lambda}^2 = \sum_{i=1}^{N-1} \frac{\{[\sigma(x) - \sigma_i]/\sigma_i\}^2}{N - \text{num}} \tag{26}$$

where σ(x) is the calculated stress that can be obtained by strain ε_i measured from an experimental test for the objective function. The unit of $\sigma(x)$ is MPa; σ_i is the stress for testing; num is the number of unknown parameters; and N is the number of experimental data observations. The optimization technique used in the present study is the genetic algorithm (GA), which can address complicated problems because it is based on the process of a biological evolution, rules governing the survival of the fittest, and the chromosome information exchange mechanism. The GA has various advantages and is well described [20,21], and Figure 4 presents a specific schematic of the GA. Table 1 lists the parameters obtained from an inverse analysis using the GA.

The stress-strain curves from the analytical solution were reproduced using the parameters obtained from inverse analysis. The test and reproduced curves are presented in Figure 5. The estimated stress-strain curve from the analytic solution was identical to the stress-strain curve from the experimental test regardless of the strain rate, and the mean square error between them was less than 10% of the strain range of the intersections. The newly derived constitutive equation and parameters from inverse analysis were validated in terms of their applicability and reliability.

3.2. Attenuation Factors in the Stress Wave Equation

The effects of the frequency of the stress wave on the attenuation in viscoelastic-plastic rock-like materials depends mainly on the imperfect elasticity (plasticity and viscosity). Previous studies have reported that microscopic factors are involved in this process, including the composition of the rock mineral, particle size, internal crack distribution, structures, and pore fluids [22]. The amplitude of attenuation is presented as:

$$A(x) = A_0 \exp(\alpha x) \tag{27}$$

where A0=A(x)|x=0 is the strength of the dynamic impact source. The attenuation amplitude of the propagating stress wave increases exponentially with increasing propagation distance, and the extent of attenuation is determined by the attenuation factor α.

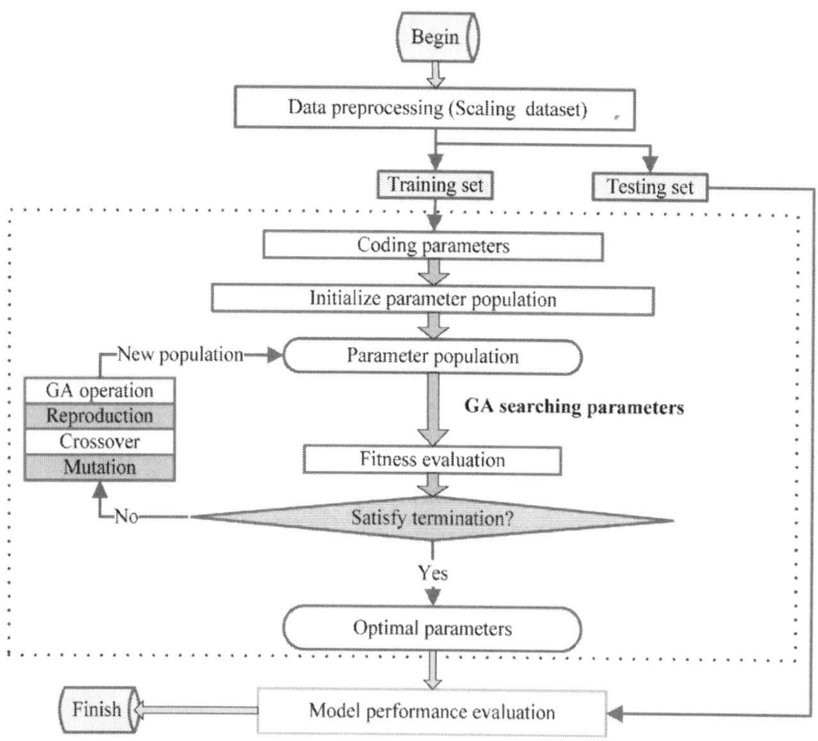

Figure 4. Schematic of the genetic algorithm (GA) used to determine parameters.

Table 1. Inverse analysis of the parameters in the constitutive equations for viscoelastic-plastic rock-like materials.

Sample Number	Mean Strain Rate (s^{-1})	Estimated Parameters from Inverse Analysis			
		Elastic Modulus (N/m^2)		Viscosity Coefficient, η	Variance, $\hat{\sigma}^2$
		E_1	E_2		
C01	87	3.499×10^{10}	3.217×10^{10}	1.200×10^4	9.077×10^{-2}
C02	194	3.498×10^{10}	3.368×10^{10}	1.093×10^4	8.572×10^{-3}
C03	260	3.488×10^{10}	3.457×10^{10}	8.311×10^3	2.239×10^{-2}

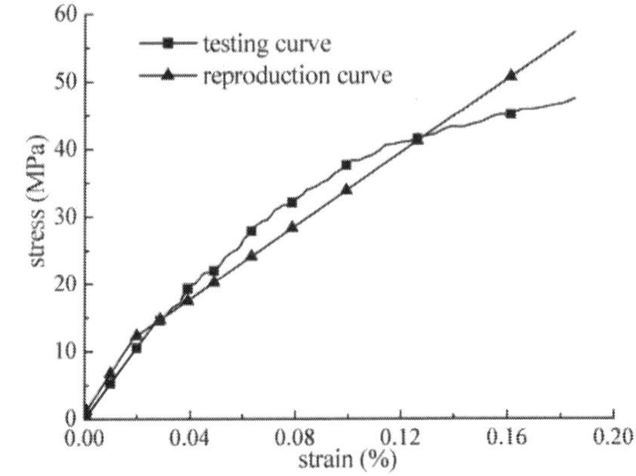

(**a**)

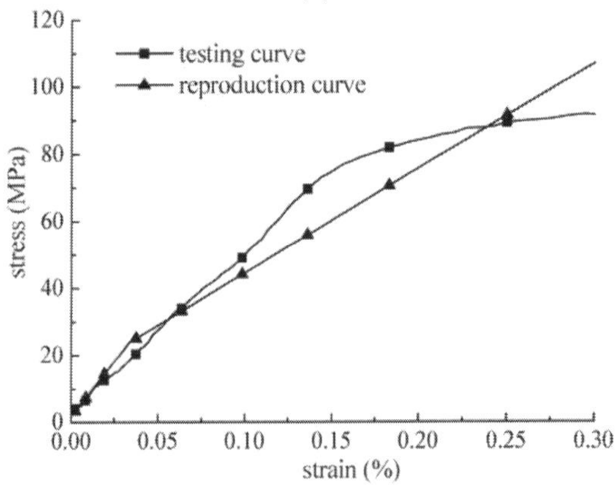

(**b**)

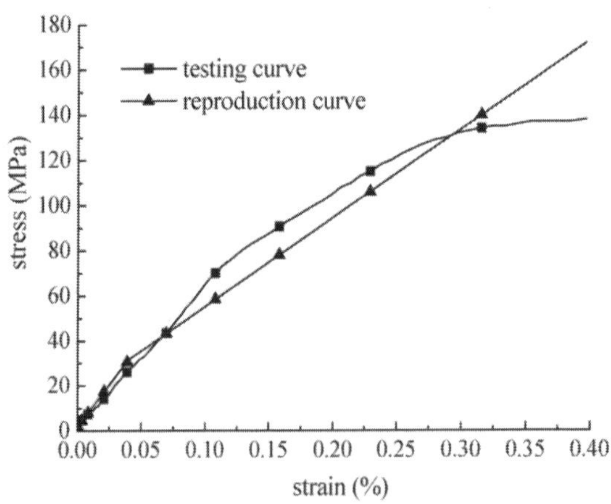

(**c**)

Figure 5. Viscoelastic-plastic dynamic stress-strain curves. (**a**) Strain rate: 87 s^{-1}; (**b**) strain rate: 194 s^{-1}; (**c**) strain rate: 260 s^{-1}.

The relationship between the attenuation factor and the frequency of the stress wave can be explored using the parameters in the constitutive equation of viscoelastic-plastic rock-like materials obtained from the experimental test using the SHPB apparatus. The viscoelasticity formula Equation (20a) and the viscoelasticity plasticity formula Equation (23a) can be used for analysis. As shown in Figure 6 and Figure 7, the square of the attenuation factor is closely correlated with the frequency of the stress wave. An increase in the frequency of the stress wave increases the attenuation factor, regardless of the strain rate.

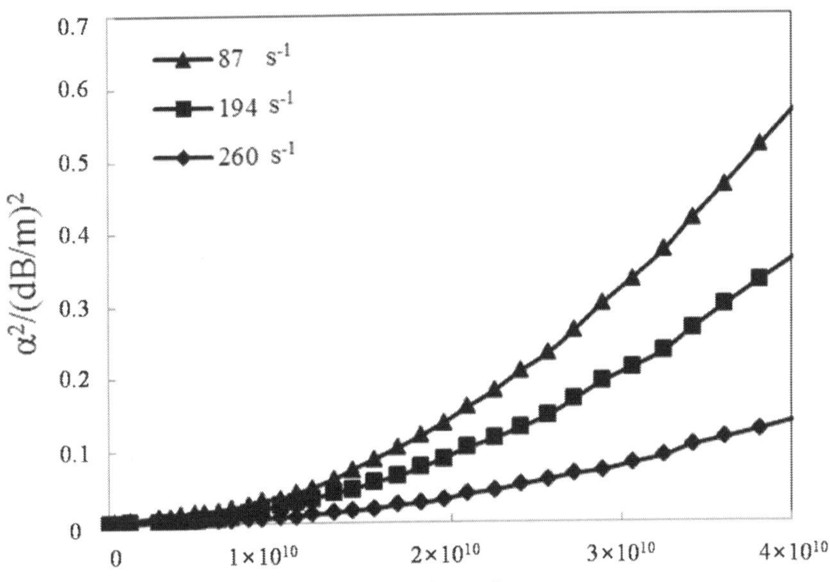

Figure 6. Effects of the stress wave frequency on attenuation factors in viscoelasticity.

According to Equation (22), the variations in the attenuation factor with respect to the elastic modulus and the viscosity coefficient were obtained, as shown in Figure 8. Based on the analysis results, the attenuation factor decreased with increasing elastic modulus, whereas it increased with increasing viscosity coefficient. On the other hand, the elastic modulus increased with increasing strain rate, and the viscosity coefficient decreased with increasing elastic modulus. The increase in the attenuation factor decelerates with increasing strain rate if the frequency of the stress wave increases. This suggests that the attenuation factor is a parameter related to the strain rate. An increase in the strain rate reduces the attenuation amplitude of the stress wave. Based on a comparison of Figure 6 and Figure 7, the elastic modulus decreases significantly based on the viscoelastic-plasticity. That is, the attenuation factor is much higher than the viscoelastic damping factor for the three strain rates.

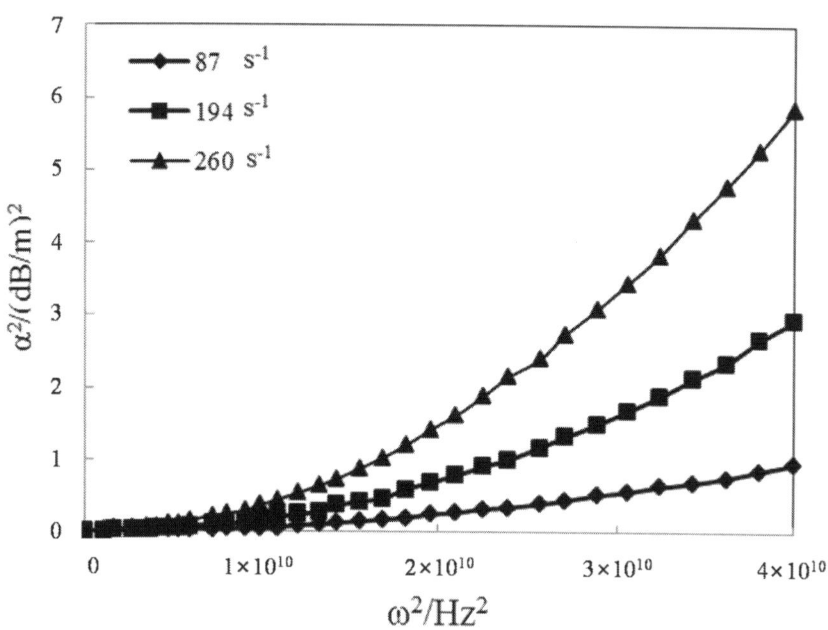

Figure 7. Effects of the stress wave frequency on the attenuation factors in viscoelastic-plasticity.

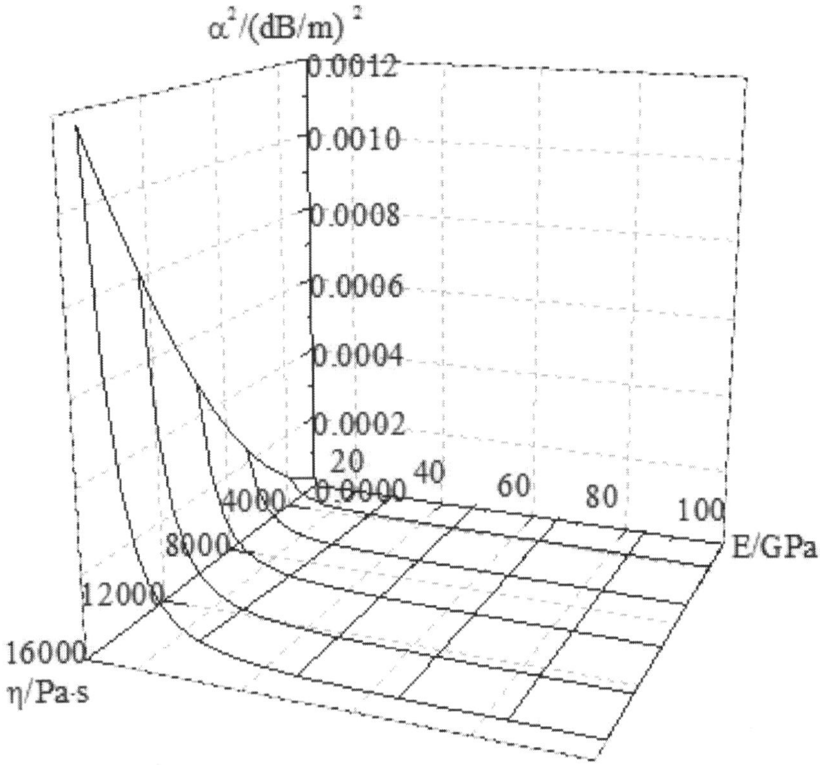

Figure 8. Variations in the attenuation factor with the elastic modulus and the viscosity coefficient.

Viscous attenuation induced by the material viscosity is called the absorption phenomenon. Compared to geometric dispersion by slender bar size, the changes in the frequency in viscous attenuation always develops in opposite directions. For the propagation of the stress wave along viscoelastic-plastic rock-like materials, the attenuation of the high-frequency wave is greater than that of the low-frequency wave, and geometric dispersion induced by the lateral inertia of the slender bar is more significant in the low-frequency wave than that in the high-frequency wave.

4. CONCLUSIONS

This theoretical and experimental study investigated the stress wave propagation in viscoelastic-plastic rock-like materials under a high-strain-rate load and examined the characteristics of strain-dependent attenuation.

Based on element model theory, composite materials, such as rock and concrete, are assumed to be a mixture composed of elastic, plastic, and viscosity components. A constitutive equation of viscoelastic-plastic rock-like materials was developed based on an analytic model composed of an elastomer, a Newtonian body, and a St. Venant body.

Stress wave equations are derived based on the kinematic and kinetic equations considering a nonlinear viscoelastic-plastic constitutive equation. The analytical solution for stress wave propagation in viscoelastic-plastic rock-like materials suggests that the attenuation factor can be determined by the viscous coefficient of the material, the elastic modulus, density, and the frequency of the stress wave.

The analytical solution derived in this study was verified through an experimental test using an SHPB apparatus. Unknown parameters of the constitutive equations were estimated inversely using a differential evolution algorithm. According to the results, the estimated stress-strain curve from the analytic solution is identical to the stress-strain curve obtained from the experimental test, regardless of the strain rate, and the mean square error between them is less than 10%.

Stress wave propagation and the attenuation factor in viscoelastic-plastic materials are affected by the viscous coefficient, the elastic modulus, density, and the frequency of the stress wave. The attenuation factor decreases with increasing elastic modulus, whereas it increases with increasing viscosity coefficient. The attenuation factor increases with increasing frequency of the stress wave. The attenuation of the stress wave is dependent on the strain such that an increase in the strain rate decelerates the increase in attenuation factor.

ACKNOWLEDGMENTS

This research was supported by the National Natural Science Foundation of China (No. 51504182), the Natural Science Basic Research Plan of Shaanxi Province of China (No. 2015JQ5187), the Scientific Research Program funded by the Shaanxi Provincial Education Department (No. 15JK1466), the Project funded by China Postdoctoral Science Foundation (No. 2015M582685), and the Xi'an University of Science and Technology Research Fund Program for Young Scholars (No. 2014-NY-018). This research was also supported by the Basic Science Research Program through the National Research Foundation (NRF) of Korea funded by the Ministry of Science, ICT, and Future Planning (No. NRF-2013R1A1A1060052). This work was supported by INHA UNIVERSITY research grant.

AUTHOR CONTRIBUTIONS

Liu Lang derived the equations and performed the experiments; Yue Zhai and Dezheng Lao analyzed the data; Hang-Lo Lee wrote and revised the paper under the supervision of KI-IL Song.

CONFLICTS OF INTEREST

The authors declare no conflict of interest.

REFERENCES

1. Follansbee, P.S.; Frantz, C. Wave propagation in the split Hopkinson pressure bar. *J. Eng. Mater. Technol.* **1983**, *105*, 61–66.

2. Li, X.B.; Lo, T.S.; Zhao, J.; Zhao, P.J. Oscillation elimination in the Hopkinson bar apparatus and resultant complete dynamic stress-strain curve for rocks. *Int. J. Rock Mech. Min. Sci.* **2000**, *37*, 1055–1060.

3. Mustaqim, M.N.M.; Song, K.I.; Cho, G.C.; Zainab, M. Long-Wavelength Elastic Wave Propagation across Naturally Fractured Rock Masses. *Rock Mech. Rock Eng.* **2014**, *47*, 561–573.

4. Zhai, Y.; Ma, G.W.; Zhao, J.H. Dynamic capability of granite and concrete under impact compressive loading. *Chin. J. Rock Mech. Eng.* **2007**, *26*, 762–768.

5. Zhai, Y.; Zhao, J.H.; Li, X.C. Study on damage viscoelasticplastic dynamic constitutive model of rock-like materials.*Chin. J. Rock Mech. Eng.* **2011**, *30*, 3820–3824.

6. Ozbolt, J.; Sharma, A. Numerical simulation of reinforced concrete beams with different shear reinforcements under dynamic impact loads. *Int. J. Impact Eng.* **2011**, *38*, 940–949.

7. Zhou, F.H.; Chen, L.; Wang, L.L. Two approaches for analyzing one-dimensional viscoelastic wave propagations. *Eng. Mech.* **2010**, *7*, 45–51.

8. Graham, G.A.C.; Sabin, G.C.W. The correspondence principle of linear viscoelasticity for problems that involve time-dependent regions. *Int. J. Eng. Sci.* **1973**, *11*, 123–140.

9. Raymond, S.; Lemiale, V.; Ibrahim, R.; Lau, R. A mesh free study of the Kalthoff-Winkler experiment in 3D at room and low temperatures under dynamic loading using viscoplasticmodeling. *Eng. Anal. Bound. Elem.* **2014**, *48*, 20–25.

10. Liu, H.S.; Tang, L.; Bo, J.S.; Feng, Z. Elastic-viscoplastic analysis of explicit FEM for simulating wave motions in rock mass. *Coal Geol. Explor.* **2009**, *37*, 38–42.

11. Passaris, E.K.S. The rheological behavior of rock salt as determined in an *in situ* pressure test cavity. In Proceedings of the 4th Conference on International Society for Rock Mechanics, Montreux, Switzerland, 2–3 September 1979; pp. 257–264.

12. Ding, J.Y.; Zhou, H.W.; Liu, D. Research on fractional derivative three elements model of salt rock. *Chin. J. Rock Mech. Eng.* **2014**, *34*, 672–678.

13. Xie, L.; Zhao, G.M.; Meng, X.R. Research on damage viscoelastic dynamic constitutive model of soft rock and concrete materials. *Chin. J. Rock Mech. Eng.* **2014**, *34*, 857–864.

14. Liu, S.; Xu, J.Y.; Chen, T.F.; Wang, P. Study on dynamic response of rock based on split Hopkinson pressure bar test.*Chin. J. Undergr. Space Eng.* **2013**, *9*, 992–995.

15. Yan, S.H.; Duan, J.X.; Yin, F.L.; Qian, Q.H. SHPB test on high-strength concrete. *J. PLA Univ. Sci. Technol.* **2000**, *1*, 6–9.

16. Zhao, H.; Gray, G. On the use of SHPB techniques to determine the dynamic behavior of materials in the range of small strains. *Int. J. Solids Struct.* **1996**, *33*, 3363–3375.

17. Hertlein, B.H. Stress wave testing of concrete: A 25-year review and a peek into the future. *Construct. Build. Mater.* **2013**, *38*, 1240–1245.

18. Wang, L.L. *Foundations of Stress Waves*; National Defense Press: Beijing, China, 2005. (In Chinese)

19. Li, Z.Q.; Chen, W.Y.; Wang, Z.H. Study on the effect of striker bar length on the SHPB measurements. *J. Mech. Strength***2010**, *32*, 942–945.

20. Liu, L.; Song, K.I.; Lao, D.Z.; Kwon, T.H. Rheological properties of cemented tailing backfill and the construction of a prediction. *Model. Mater.* **2015**, *8*, 2076–2092.

21. Zhou, J.; Li, X.B.; Shi, X.Z. Long-term prediction model of rockburst in underground openings using heuristic algorithms and support vector machines. *Saf. Sci.* **2012**, *50*, 629–644.

22. Li, J.G.; Cheng, G.Q.; Zou, Y.H. Study on the Stress Wave Propagation Attenuation Laws in the Elastoplastic Coal and Rock. Master's Thesis, Shandong University of Science and Technology, Qingdao, China, 2008. pp. 20–60.

CHAPTER 7

Towards a Mathematical Model for Elastic Wave Propagation in Granular Materials

Leonardo Trujillo[1,2*], Vanessa Torres[1], Franklin Peniche[1,3], Leonardo Di G. Sigalotti[1]

[1] Centro de Física, Instituto Venezolano de Investigaciones Científicas (IVIC), Caracas, Venezuela

[2] The Abdus Salam, International Centre for Theoretical Physics (ICTP), Trieste, Italy

[3] Departamento de Física y Electrónica, Universidad de Cordoba, Montería, Colombia

ABSTRACT

A theoretical model for the propagation of acoustic waves in dry granular media is presented within the framework of the nonlinear granular elasticity. An essential ingredient is the dependence of the elastic moduli on compression. For the purpose of illustration, we analyze the case of a time-harmonic plane wave propagation under isotropic compression. We derive explicit relations for the wave speed dependence with the confining pressure. The present approach provides an accurate description of acoustic wave propagation in granular packings and represents a powerful tool to interpret the results of current experiments.

KEYWORDS

Granular Media; Elastic Waves; Nonlinear Elasticity; Hydrostatic Compression; Granular Elasticity

1. INTRODUCTION

Granular materials consist of a collection of discrete macroscopic solid particles interacting via repulsive contact forces. Perhaps the simplest example of such a system is a dense packing of spherical glass beads under hydrostatic confining pressure. Such model systems are a useful starting point in the description of composite materials. Their physical behaviour involves complex nonlinear phenomena, such as non equilibrium configurations, energy dissipation and nonlinear elastic response [1]. We know from classical continuum dynamics that when a deformable body is under the action of a force distributed over its surface, the force is transmitted to every point inside the body [2]. Conversely, when we deal with a static packing of granular particles, the way by which the forces are transmitted within the packing remains a complex, and still unresolved, problem [3]. Several features of granular matter may be responsible for the complexity of the response to external perturbations. One important aspect lies on the observation that force networks form the skeleton that carries most of the load in a static granular medium [4]. The mechanical response of granular packings to external perturbations plays a major role in numerous engineering endeavours (such as soil mechanics and geophysics), as well as in industry (oil exploration, structural stability, product formulation in pharmacology and domestics, granular composites, heterogeneous materials, etc.).

Most laboratory experiments have been carried out using two-dimensional packings, consisting of photoelastic (i.e., birefringent under strain) disks, which have then allowed for a finest visualization of the generation and dynamical evolution of force chains. However, real granular materials are optically opaque, the photoelastic technique becomes difficult to practice. On the other hand, new tools such as pulsed ultrasonic transmission through granular beds under confined stress have been recently deviced to understand the elastic response of three-dimensional granular packings [5-7]. In particular, by studying the low-amplitude coherent wave propagation and multiple ultrasound scattering, it is possible to infer many fundamental properties of granular materials such as elastic constants and dissipation mechanisms [8].

The understanding of wave motion in granular media took a major step forward with the experimental observation of the coexistence of a coherent ballistic pulse travelling through an effective contact medium and a multiply scattered signal [5]. Such a picture is confirmed by the experiments of sound propagation in two dimensional regular lattice of spheres under isotropic compression [9], where the transmitted coherent signal remains almost unchanged for different packing realizations followed by an incoherent tail which depends on the specific packing configuration, in agreement with the completely disordered three dimensional case [5,6]. Besides the sensitive configuration coda type high frequency (HF) incoherent signals, one observes a primary low frequency (LF) coherent pulse [6,9]. This coherent pulse well defined at the leading edge of the transmitted signal corresponds to an effective wave propagating ballistically at compressional velocity $c \sim 1000$ m/s, and frequency $f \sim 70$ kHz. The sensitivity of the coherent and incoherent waves to changes in packing configurations has been studied in [6]. For independent granular samples, an ensemble average of configurations can cancel scattered signals and leave only the coherent wave. From experiments [6] one can infers that the coherent pulse is a self-averaging signal and configuration insensitive, which thus probes sound propagation in an equivalent effective contact medium.

Usually the elastic response of a compressed granular packing is modeled as an assembly of spherical particles, where the contact forces between them are described by the Hertz law [10]. This model predicts that the speed of sound scales as 1/6 with the confining pressure P. However, experiments [6] (3D disordered packing) and [9] (2D regular lattice) show that for moderate pressures the speed of sound scales with a non-Hertzian exponent, roughly 1/4, followed by a Hertzian one at large values of the confining pressure.

In this paper we deal with a continuum description of wave propagation in granular packings, using the nonlinear elastic model developed in Refs [11-13]. This paper is structured as follows: In Section 2 we present a concise and detailed pedagogical introduction to granular elasticity; In Section 3 we introduce the equation of motion to describe the dynamical response of a granular packing subjected to an external perturbation and it is verified that the theory is consistent with classical elasticity; In Section 4 we solve the equation of motion for time-harmonic plane wave propagation under hydrostatic compression. Then, we derive a mathematical expression for the longitudinal wave speed with pressure dependence. In Section 5 we summarize the relevant conclusions.

2. GRANULAR ELASTICITY

A basis for the construction of a theory for elastic wave propagation in granular media is provided by the elastic theory conveniently modified to account in infinitesimal forces due to force-dependent internal contacts between grains. The goal of this section is to introduce the granular elastic theory proposed by Jiang and Liu [11,12], which emphasizes the role of intrinsic features of granular dynamics such as volume dilatancy, mechanical yield and anisotropies in the stress distribution. Another fundamental ingredient is that the theory takes into account a general form of the type of contacts between the grains. On the other hand, the formulation directly extends the Boussinesq modified theory of elasticity and removes their thermodynamic inconsistency via a correct formulation of the strain free energy functional. In addition, under certain limits the theory includes the wellknown linear elasticity of isotropic and homogeneous elastic solids.

We begin with the basic notions used in the construction of a self-contained elastic theory for granular materials. To illustrate how a granular packing can be described by an elastic approach, we make the following assumptions:

- As a very first rule, we impose that the theory we will build should be invariant with respect to any Galilean transformation, i.e., any translation or rotation or combination of both, added to the displacement field of the medium should neither change their energy, nor any physical quantity we can derive from it such as the stress field.

- We do not consider the case for plastic deformations in the granular packing. Therefore, we assume that the predominant part of the energy is elastic. Elasticity implies reversible deformation, which change the internal energy ε of the deformed body according to $d\varepsilon = Tds + dW$, where T is the (thermal) temperature, S is the entropy and W the work due to the force exerted on the packing. We assume a reference equilibrium state a temperature T, from which we can construct an elastic energy potential (free energy functional) F. For a static granular media the granular temperature $T_g = 0$, which "measures" kinetic energy fluctuations when the system is fluidized. For a more detailed discussion and clarifications see references [11-13].

- As a mechanical model for the granular particles we consider smooth hard spheres with diameter d. We do not include friction at the contact surface between the spheres, and rotational degrees of freedom are neglected. Therefore, we consider the potential energy as a path-independent function.

- Congruently with the previous point, distributions of body or surface couples are not include in the present analysis. In this sense the present formulation is not complete and must be extended to include the effects of the grain rotation into the corresponding constitutive relations.

2.1. Continuum Theory Framework

The analysis of the deformation of a granular packing is rendered amenable to mathematical analysis by introducing the concept of a continuum medium. In this idealization it is assumed that properties averaged over a mesoscale are continuous functions of position and time. However, the presence of force chains at the grain scale, implying preferred force paths, have served as empirical argument against an isotropic continuum description of granular matter [3,4]. Nonetheless, recent findings on the stress distribution response to local and global perturbations have shed some light on the validity of using a continuum theory [14,15]. The passage from a microscopic to a macroscopic (continuum) mechanical description of granular and heterogeneous materials, including the mesoscopic disorder, has been recently addressed by Goldenberg and Goldhirsch [16-19]. They showed that exact continuum forms of the balance equations (for mass, momentum and energy) can be established as relations between spatially-weighted sums, in which a physical quantity $F(r)$ is approximated by the summation interpolant $F(r) \approx \sum_i F_i(r) W(r - r'|h)$, where $W(r - r'|h)$ is a smooth (differentiable) function, commonly referred as the interpolating kernel, and h is the smoothing (coarse-graining) length, which determines the spatial resolution and the scales for the spatial averages. These findings justify the continuum analysis adopted in the present work. In passing, we note that the approach followed in references [16-19] seems to be congruent with the Smoothed Particle Hydrodynamics (SPH) computational method; a meshfree particle method based on Lagrangian formulation, which has been widely applied to different areas in engineering and science [20].

2.2. Deformation: Displacements and Strains

The displacement field inside the granular packing will be characterized by a vector field u(r). Moreover, we require that the theory starts with assumption of small deformations, i.e., for an infinitesimal deformation the displacements field and their gradients are small compared to unity $(|u| \text{ and } |\nabla u| \ll 1)$. This choice limit the range of applicability of the theory but is physically reasonable for granular systems under strong static compression and small amplitude wave motion. However, the present model could provides the theoretical basis for analysis of ultrasonic wave propagation in geotechnical engineering, soil mechanics and rock mechanics.

To comply with the requirement of translational invariance, we express the elastic energy of the granular packing as a function of the derivatives of the displacement field. Since the displacement is a vector field, the gradient is a second order tensor whose components i; j can be written $\partial u_j / \partial x_i$. We can split the gradient into a symmetric and antisymmetric tensor. The later will contain the rotational part of the field. Because of invariance under rotations, we require that the energy should not depend on the antisymmetric part of the gradient of displacement. On the other hand, the displacement associated with deforming the grains, stores energy reversible and maintains a static strain. Sliding and rolling lead to irreversible, plastic process that only heat up the system. So, the total strain tensor u_{ij} may be decomposed into elastic and plastic parts $u_{ij} = c_{ij} + c_{ij}^{p}$. In this work we limit our analysis to granular packings under strong static compression. Therefore, the granular energy is a function of the elastic energy alone and we can neglect the plastic strain contribution. Up to linear order, the strain tensor is

$$u_{ij} \approx \epsilon_{ij} = \frac{1}{2}\left(\frac{\partial u_j}{\partial x_i} + \frac{\partial u_i}{\partial x_j}\right).$$

Let us remark that in the present analysis we do not include non-affine deformations. This assumption should be valid for strong compressed granular packings far from the jamming transition [21].

2.3. Strain Energy

In order to have an intrinsic form for the energy, i.e. which does not depend on the basis chosen to express the displacement field, we need to introduce only invariants of the tensor $\hat{u}$. A second order tensor in three dimensions, has only three invariants:

$$u_I = u_{ij} = \mathrm{Tr}\left(\hat{u}\right); \quad u_{II} = u_{ij}\, u_{ji} = \mathrm{Tr}\left(\hat{u}^2\right);$$

and $u_{III} = u_{ij}\, u_{jl}\, u_{il} = \mathrm{Tr}\left(\hat{u}^3\right)$.

Here $\mathrm{Tr}(\cdot)$ denotes the trace. In order to build a linear theory, we have to introduce an expression which is quadratic in the strain field. Thus only the square of the first invariant, and the second invariant, together with the assumptions of Galilean invariance and thermodynamic equilibrium, can be used and we are naturally led to the general expression of the elastic potential energy, given by the Helmholtz free energy functional [2]:

$F(\hat{u}) = \frac{1}{2}\left[\lambda \mathrm{Tr}(\hat{u})^2 + 2\mu \mathrm{Tr}(\hat{u}^2)\right]$ where λ and μ are material-dependent constants which characterizes the rigidity of the solids and are called the Lamé coefficients. If $\hat{u}$ contains only off-diagonal components, like for instance a pure shear, the elastic constant that comes into play is $\mu = \mathrm{G}$, the shear modulus. In the case of an isotropic compression, $\hat{u}$ is proportional to the identity tensor. The elastic constant relating the pressure to the decrease of volume is now $K = \lambda + 2/3\mu$, the compressibility modulus. In a model for granular materials, the elastic moduli K and G must be assumed to be proportional to the volume compression $\Delta = 1 - \rho_0/\rho = -u_{ll}$, where ρ_0 is the density in the absence of external forces [11,12]. So we have

$$K = \tilde{K}\Delta^b, \quad G = \tilde{G}\Delta^a, \tag{1}$$

with $\tilde{K}, \tilde{G} > 0$ for $\Delta \geq 0$ and $\tilde{K}, \tilde{G} = 0$ for $\Delta = 0$, so that the elastic moduli remain finite. The exponents a and b are related to the type of contact between the grains. That is, when $a = b = 0$ linear elasticity is recovered, whereas $a = b = 1/2$ implies Hertz contacts [10]. This formulation provides a much better approximation to granular elastic behavior in which we can specify the type of contact by suitably choosing the exponents a and b. The Helmholtz free energy in the Jiang-Liu model is [11,12]:

$$F_{JL} = \frac{1}{2}\left[\tilde{K}\Delta^b\right]u_{ll}^2 + \left[\tilde{G}\Delta^a\right]u_s^2, \tag{2}$$

where $u_s^2 = u_{ij}^0 u_{ij}^0$, with $u_{ij}^0 = u_{ij} - 1/2 u_{ll}\delta_{ij}$.

The free energy strain potential (2) is stable only in the range of strain values that keeps it convex. Therefore, Equation (2) naturally accounts for unstable configurations of the system, as the yield, which appears as a phase transition on a potential-strain diagram.

2.4. Stress Field and Equilibrium

We now introduce the conjugated field to the strain: the granular stress field. By definition, this stress field, which is also a second-order symmetric tensor field, σ_{ij} is related to the strain field by

$$\sigma_{ij} = \frac{\partial F_{JL}}{\partial u_{ij}},$$

$$= -Ku_{ll}\delta_{ij} + 2Gu_{ij}^{0} - \frac{1}{\Delta}\left(\frac{1}{2}bKu_{ll}^{2} + aGu_{s}^{2}\right) \quad (3)$$

where δ_{ij} is the Kronecker's delta. Galilean invariance implies that the stress field should not do any work during a rigid motion of the solid. This property can be used to derive the balance equation satisfied by the stress field, always valid even if the rheological behaviour considered is not linear. When the granular packing is subjected to an external force density f, the balance equations read

$$\nabla \cdot \hat{\sigma} + f = 0. \quad (4)$$

This equation have been checked in references [22,23], where the authors analyzed several features of static granular packings, such as the stress dip in sand piles, the stress distribution in silos and under point loads, and the variation with pressure of the elastic moduli.

3. GRANULAR ELASTODYNAMICS

Now we are ready to formulate a mesoscopic elastodynamic theory for granular materials. Above we stated that the theory is founded on the assumption of a continuum (coarse grained) field description and an elastic theory for granular materials which takes into account the compression dependent elastic moduli.

Equation of Motion

We start with the equation of motion for the elastic displacement u at time t and postion r,

$$\rho \frac{\partial^2}{\partial t^2} u(r,t) = \nabla \cdot \hat{\sigma} + f(r,t). \tag{5}$$

From the Jiang-Liu elastic model the stress tensor (3) which, by the Hooke's law, is given in terms of the Lamé coefficients by

$$\hat{\sigma} = -\left\{\left[1+\frac{b}{2}\right]\lambda + \left[\frac{b}{3}+a\frac{u_s^2}{\Delta^2}\right]\mu\right\}(\nabla \cdot u)\hat{I} \\ -\mu\left[\nabla u + (\nabla u)^T\right] \tag{6}$$

where $\hat{I}$ denotes the unitary tensor and the superindex ^{T} means transposition. Inserting Equation (6) into Equation (5) and rearranging terms, the equation of motion is

$$\rho \frac{\partial^2}{\partial t^2} u(r,t) = \left\{\left[1+\frac{b}{2}\right]\lambda \right. \\ \left. +\left[2+\frac{b}{3}+a\frac{u_s^2}{\Delta^2}\right]\mu\right\}\nabla\left[\nabla \cdot u(r,t)\right] \\ -\mu\nabla\times\nabla\times u(r,t) + f(r,t), \tag{7}$$

this expression indicates that the elastic wave have both dilatational $\nabla \cdot u(r,t)$ and rotational deformations $\nabla \times u(r,t)$. Let us remark that setting $a=b=0$, as it would be for linear elasticity, Equation (7) reduces to

$$\rho\frac{\partial^2}{\partial t^2}u(r,t)=(\lambda+2\mu)\nabla[\nabla\cdot u(r,t)] \\ -\mu\nabla\times\nabla\times u(r,t)+f(r,t). \tag{8}$$

The above equation further simplifies to the wellknown wave equation

$$\frac{\partial^2}{\partial t^2}u(r,t)-c_p^2\nabla[\nabla\cdot u(r,t)] \\ -c_s^2\nabla\times\nabla\times u(r,t)=\rho^{-1}f(r,t) \tag{9}$$

The terms $c_p\sqrt{(\lambda+2\mu)/\rho}$ and $c_s=\sqrt{\mu/\rho}$ are the compressional wave speed, and shear or transverse wave speed, respectively. This proves that the present elastodynamic model includes the well-known linear elasticity of isotropic and homogeneous elastic solids.

4. WAVE PROPAGATION

4.1. Time-Harmonic Plane Wave Propagation under Hydrostatic Compression

Now we proceed to analyze the case of time-harmonic plane wave under compression. We shall assume a perturbation of the packing by a time-harmonic plane deformation given by

$$u^P=U\exp[i(k\cdot r-\omega t)], \tag{10}$$

where the vector U is the amplitude of the displacement along the plane of the wave (polarization), while k is the wave vector and ω is the pulsation. We will now consider the case for isotropic compression. In the unperturbed reference configuration the strain tensor is

$u_{ij} = u_{ll}\delta_{ij}$, i.e.,

$$\hat{u} = \begin{pmatrix} u_{11} & 0 & 0 \\ 0 & u_{22} & 0 \\ 0 & 0 & u_{33} \end{pmatrix}, \tag{11}$$

and the trace is $\mathrm{Tr}(u_{ij}) = (u_{11} + u_{22} + u_{33}) = -\Delta$. The traceless part, $u_{ij}^0 = u_{ij} + 1/3\Delta\delta_{ij}$, is

$$\hat{u}^0 = \begin{pmatrix} 2u_{11} - u_{22} - u_{33} & 0 & 0 \\ 0 & -u_{11} + 2u_{22} - u_{33} & 0 \\ 0 & 0 & -u_{11} - u_{22} + 2u_{33} \end{pmatrix}. \tag{12}$$

Therefore, the invariant u_s^2 is

$$u_s^2 = \frac{2}{3}\Delta - 2(u_{11}u_{22} + u_{11}u_{33} + u_{22}u_{33}). \tag{13}$$

Finally, from (3) the stress tensor is

$$\sigma_{ij} = \left[\left(1 + \frac{b}{2}\right)K + a\frac{u_s^2}{\Delta^2}G\right]\Delta\delta_{ij} - 2u_{ij}^0 G. \tag{14}$$

Now we consider the displacements parallel to the wave normal plane along the x_1 axis and amplitude $\tilde{u}$

$$\tilde{u}_{\|}^{P} = \begin{pmatrix} \tilde{u} \\ 0 \\ 0 \end{pmatrix} \exp\left[\mathrm{i}\left(k_1 x_1 - \omega t\right)\right]. \tag{15}$$

As a next step, we need to calculate the disturbances. Thus the strain tensor associated with the wave is

$$\hat{u} = \mathrm{i}k\xi \begin{pmatrix} 1 & 0 & 0 \\ 0 & 0 & 0 \\ 0 & 0 & 0 \end{pmatrix}, \tag{16}$$

where $\xi = \tilde{u}\exp\left[\mathrm{i}\left(k_1 x_1 - \omega t\right)\right]$. The volume compression is $\tilde{\Delta} = -\mathrm{i}k\xi$. On the other hand, the traceless part $\tilde{u}_{ij}^{0} = \tilde{u}_{ij} - \frac{1}{3}\tilde{u}_{ll}\delta_{ij}$ is

$$\hat{\tilde{u}}^{0} = \frac{1}{3}\mathrm{i}k\xi \begin{pmatrix} 2 & 0 & 0 \\ 0 & -1 & 0 \\ 0 & 0 & -1 \end{pmatrix}. \tag{17}$$

Now the invariant $\tilde{u}_s^2$ is

$$\tilde{u}_s^2 = \frac{2}{3}\mathrm{i}k\xi\left(4u_{11} + \Delta\right). \tag{18}$$

The components of the perturbed stress tensor are given by

$$\tilde{\sigma}_{ij} = \left[(1+b)\left(1+\frac{b}{2}\right)K - a(1-a)\frac{u_s^2}{\Delta^2}G\right]\tilde{\Delta}\delta_{ij} + a\frac{\tilde{u}_s^2}{\Delta}G\delta_{ij} - 2\left[\tilde{u}_{ij}^{0} + a\frac{u_{ij}^{0}}{\Delta}\tilde{\Delta}\right]G, \tag{19}$$

which in matrix notation have the diagonal form $\hat{\tilde{\sigma}}_{ij} = \mathrm{diag}\left(\sigma_{11}, \sigma_{22}, \sigma_{33}\right)$, where

$$\sigma_{11}=\left[(1+b)\left(1+\frac{b}{2}\right)K-\left(a(1-a)\frac{u_s^2}{\Delta^2}\right.\right.$$
$$\left.\left.-a\frac{\tilde{u}_s^2}{\Delta\tilde{\Delta}}+\frac{2}{3}\frac{a}{\Delta}(2u_{11}-u_{22}-u_{33})-\frac{4}{3}\right)G\right]\tilde{\Delta},\tag{20}$$

$$\sigma_{22}=\left[(1+b)\left(1+\frac{b}{2}\right)K-\left(a(1-a)\frac{u_s^2}{\Delta^2}\right.\right.$$
$$\left.\left.-a\frac{\tilde{u}_s^2}{\Delta\tilde{\Delta}}+\frac{2}{3}\frac{a}{\Delta}(2u_{22}-u_{11}-u_{33})+\frac{2}{3}\right)G\right]\tilde{\Delta},\tag{21}$$

$$\sigma_{33}=\left[(1+b)\left(1+\frac{b}{2}\right)K-\left(a(1-a)\frac{u_s^2}{\Delta^2}\right.\right.$$
$$\left.\left.-a\frac{\tilde{u}_s^2}{\Delta\tilde{\Delta}}+\frac{2}{3}\frac{a}{\Delta}(2u_{33}-u_{11}-u_{22})+\frac{2}{3}\right)G\right]\tilde{\Delta}.\tag{22}$$

We can verify that $\partial\sigma_{22}/\partial x_2=\partial\sigma_{33}/\partial x_3=0$. Finally, using Equation (15) and (20) we can ensemble the equation of motion to obtain

$$\rho\omega^2\xi=\left[(1+b)\left(1+\frac{b}{2}\right)K-\left(a(1-a)\frac{u_s^2}{\Delta^2}-a\frac{\tilde{u}_s^2}{\Delta\tilde{\Delta}}\right.\right.$$
$$\left.\left.+\frac{2}{3}\frac{a}{\Delta}(2u_{11}-u_{22}-u_{33})-\frac{4}{3}\right)G\right]k_1^2\xi.\tag{23}$$

As we are interested in the case for isotropic compression, the components for the strain are $u_{11}=u_{22}=u_{33}=1/3u_0$, thus $u_s^2=\tilde{u}_s^2=0$. Therefore, the resulting equation of motion is

$$\rho\omega^2\xi=\left[(1+b)\left(1+\frac{b}{2}\right)K+\frac{4}{3}G\right]k_1^2\xi.\tag{24}$$

The pulsation ω and the wave vector k_1 define the longitudinal wave speed, i.e.,

$$v_p := \frac{\omega}{k_1}. \tag{25}$$

Furthermore, noting that $K = \tilde{K}\Delta^b$ and $G = \tilde{G}\Delta^a$, the longitudinal wave speed is

$$v_p = \sqrt{\frac{1}{\rho}}\left[(1+b)\left(1+\frac{b}{2}\right)\tilde{K}\Delta^b + \frac{4}{3}\tilde{G}\Delta^a\right]^{1/2}. \tag{26}$$

4.2. Speed of Sound and Pressure Dependence

In Refs. [5] and [9] the dependence of v_p vs the applied external pressure P was analyzed. The experiments show a scaling dependence for v_p with P. In order to capture the dependence for the calculated longitudinal wave speed (26) with the pressure, we use the granular free-energy functional (2). The pressure is given by

$$P = \frac{\partial F_{JL}}{\partial \Delta}, \\ = \left(1+\frac{b}{2}\right)\tilde{K}\Delta^{b+1} \tag{27}$$

Therefore, the compression Δ is:

$$\Delta = \left[\frac{1}{\left(1+\frac{b}{2}\right)}\right]^{\frac{1}{b+1}}\left(\frac{P}{\tilde{K}}\right)^{\frac{1}{b+1}}. \tag{28}$$

If $a = b$, Equation (26) can be written as

$$\mathrm{v}_p = c_p \left(\frac{P}{\tilde{K}} \right)^{\frac{b}{2(b+1)}}, \tag{29}$$

where c_p is defined as

$$c_p = \left[(1+b)\left(1+\frac{b}{2}\right)\frac{\tilde{K}}{\rho} + \frac{4}{3}\frac{\tilde{G}}{\rho} \right]^{1/2} \left[\frac{1}{\left(1+\frac{b}{2}\right)} \right]^{\frac{b}{2(b+1)}}. \tag{30}$$

The present theoretical analysis, let us to conclude that the longitudinal wave speed scales with the pressure as

$$\mathrm{v}_p \sim P^{\frac{b}{2(b+1)}}, \tag{31}$$

where b is given by the mechanics of the contact model.

According to experiments [5] and [9] we have the following two cases:

For $b = 1/2$ (Hertzian contacts):

$$\begin{aligned} \mathrm{v}_p(P)\big|_{b=\frac{1}{2}} &= 0.963\left[\frac{15}{8}\frac{\tilde{K}}{\rho_0} + \frac{4}{3}\frac{\tilde{G}}{\rho_0} \right]^{1/2} \left(\frac{P}{\tilde{K}} \right)^{\frac{1}{6}} \\ &= c_p^H \left(\frac{P}{\tilde{K}} \right)^{\frac{1}{6}} \sim P^{\frac{1}{6}} \end{aligned} \tag{32}$$

For $b = 1$ (non-Hertzian contacts):

$$v_p(P)\big|_{b=1} = 0.904\left[3\frac{\tilde{K}}{\rho_0}+\frac{4}{3}\frac{\tilde{G}}{\rho_0}\right]^{1/2}\left(\frac{P}{\tilde{K}}\right)^{\frac{1}{4}}$$
$$= c_p^{nH}\left(\frac{P}{\tilde{K}}\right)^{\frac{1}{4}} \sim P^{\frac{1}{4}} \qquad (33)$$

Equations (32) and (33) behave according to the experiments in a qualitative manner. In **Figure 1**we compare the above equations with the data reported in [5].

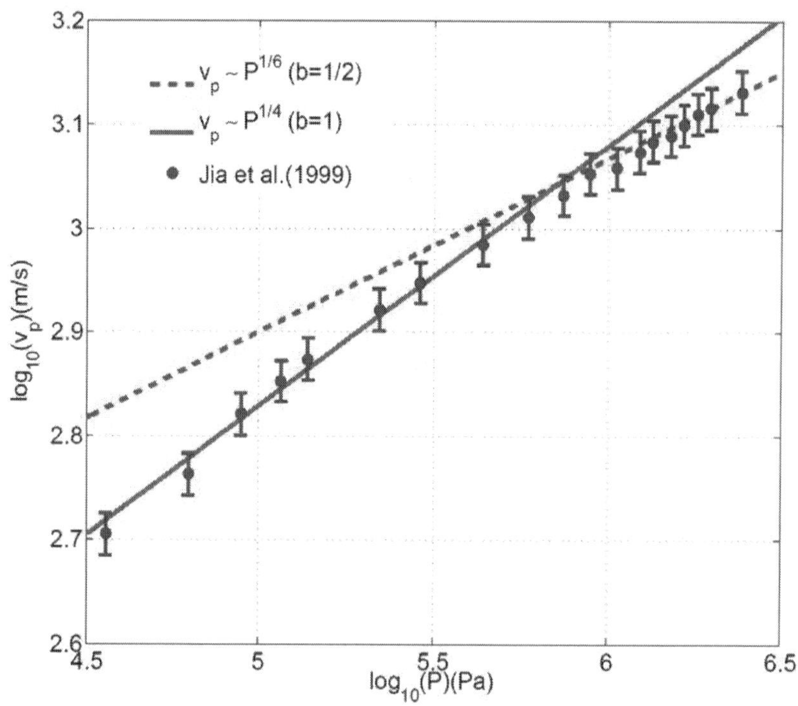

Figure 1. Velocity sound speed v_p (data points [5]) of the coherent wave versus the applied stress P. Red line: b = 1/2 (Hertzian contacts) with c_p^H = 7.9; Green dashed line: b = 1 with c_p^{nH} = 4.9.

However, at this point we have not a criterium to understand why the velocity "changes" with P between Equations (32) and (33), i.e., from the non-Hertzian contact model $b=1$ (33) for moderate pressures, to the Hertzian contact model $b=1/2$ (32) for higher pressures.

From the experimental data one can characterize the pressure dependence of the sound velocity introducing an effective exponent defined by the logarithmic derivative between v_p and P

$$\nu = \frac{d \ln v_p}{d \ln P}. \tag{34}$$

The compression Δ clearly influences the behavior of the velocity because it depends on the pressure. From Equation (29) we can estimate the dependence on Δ as

$$\left(\frac{v_p}{c_p}\right)^{\frac{2}{b}} = \Delta\left[\frac{1}{\left(1+\frac{b}{2}\right)}\right]^{-\frac{1}{b+1}} = \left(\frac{P}{\tilde{K}}\right)^{\frac{1}{b+1}}. \tag{35}$$

The qualitative behavior of ν can be obtained by interpolating from the asymptotic behavior at large and moderate pressures. This opens the question about if one can assume that ν is continuos, monotonically decreasing function from $\nu=1/4$ to $\nu=1/6$, while P increases. Or if it corresponds to discontinuous transition between the exponents.

Now we are in position to relate the parameter b and the exponent ν. We can estimate that b behaves according to the effective exponent ν as

$$b \sim \frac{\nu}{\frac{1}{2}-\nu}. \tag{36}$$

Verifying from the experiments, that the high pressure limit $\nu = 1/6$, implies for our model that $b \sim 1/2$. While for moderate pressures $\nu \sim 1/4 \Rightarrow b \sim 1$. The first case clearly corresponds to the Hertz contact model in the Jiang-Liu theory. The resistance to an external compression by a granular packing is due macroscopic rigidity mediated by the small contact regions between the grains and the contact areas increase with pressure. For spheroidal objects the Hertz model implies that the amount of deformation h increases with P as $h \sim P^{\frac{2}{3}}$ and the elastic energy U_H of the deformed sphere behave as $U_H \sim h^{\frac{5}{2}}$. This corresponds to the scaling at the high pressure regime. Various interpretations have been proposed for non-Hertzian contacts (e.g., [24,25]). However, let us remark that in 1971 Johnson, Kendal and Roberts [26] quoted some experimental outcomes where for moderated loads contacts areas between the bodies were considerably larger than those predicted by the Hertz model, while for high loads the results closely fitted the Hertz theory. Their observation is congruent with the scaling transition reported in [5] and [9]. The JKR theory [26] introduces the effects due to the energy at the surfaces of the contact bodies. According to [26], at moderate pressures tensile forces are present at the deformed surface modifying the Hertz contact law. They proposed that the tensile forces act as an attractive surface force even for clean and dry surfaces. These surfaces forces are of little significance at high loads, but become more significative as the pressure is reduced.

5. CONCLUSIONS

In this paper, we have shown that the nonlinear elastic theory proposed by Jiang and Liu [11] can be used to derive a time-evolution equation for the displacement field. The present formulation naturally accounts for features that are intrinsic to the dynamics of granular matter, considering microstructure information through the type of contact between the grains. The theory shows that the wave speed scales with pressure as $v_p \sim P^{\frac{b}{2(b+1)}}$. From the experimental data we have introduced the effective exponent υ and their relationship with the parameter b. This was done to characterize the transition from the non-Herztian regime $P^{1/4}$ to the Herztian regime $P^{1/6}$. However, we can not conclude if υ is a continuos, monotonically decreasing function of P, or a discontinuous transition at some critical value of P_c given by the intersection between Equations (32) and (33). On the other hand, we have provided an interpretation for the non-Hertzian contacts based on the JKR theory [26]. We hope that the present theoretical findings will stimulate further experimental works to explore in detail the behavior of contact deformations under compression and the effects on wave propagation in granular materials.

The mathematical model reported here is a first step towards the implementation and validation of computational engineering simulations for a variety of problems related to elastic wave propagation in granular composites. The equations of motion (7) are three partial differential equations which can be solved numerically for u_1, u_2 and u_3 provided with appropriate boundary and initial conditions, as for example using finite-element codes. In addition, the mathematical formulation at hand could help to inspire new ideas towards the construction of working theories for granular media, which may ultimately improve our interpretation of present experiments. For instance, understanding how waves do actually propagate in heterogeneous media is relevant to well-known problems such as earthquakes and seismic wave propagation and reflection in order to estimate the hydrocarbon content of potential oil wells.

REFERENCES

1. H. M. Jaeger, S. R. Nagel and R. P. Behringer, "Granular Solids, Liquids, and Gases," Review of Modern Physics, Vol. 68, No. 4, 1996, pp. 1259-1273.

2. L. D. Landau and E. M. Lifshitz, "Theory of Elasticity," Pergamon Press, New York, 1970.

3. S. Luding, "Granular Media: Information Propagation," Nature, Vol. 435, No. 7039, 2005, pp. 159-160.

4. T. S. Majmudar and R. P. Behringer, "Contact Force Measurements and Stress-Induced Anisotropy in Granular Materials," Nature, Vol. 435, No. 7045, 2005, pp. 1079- 1082.

5. X. Jia, C. Caroli and B. Velicky, "Ultrasound Propagation in Externally Stressed Granular Media," Physical Review Letters, Vol. 82, No. 9, 1999, pp. 1863-1866.

6. X. Jia, "Codalike Multiple Scattering of Elastic Waves in Dense Granular Media," Physical Review Letters, Vol. 93, No. 15, 2004, pp. 154303: 1-4.

7. X. Jia, J. Laurent, Y. Khidas and V. Langlois, "Sound Scattering in Dense Granular Media," Chinese Science Bulletin, Vol. 54, No. 23, 2009, pp. 4327-4336.

8. T. Brunet, X. Jia and P. Mills, "Mechanisms for Acoustic Absorption in Dry and Weakly Wet Granular Media," Physical Review Letters, Vol. 101, No. 13, 2008, pp. 138001: 1-4.

9. B. Gilles and C. Coste, "Low-Frequency Behavior of Beads Constrained on a Lattice," Physical Review Letters, Vol. 90, No. 17, 2003, pp. 174302: 1-4.

10. H. Hertz, "Über die Berührung Fester Elastischer Körper," Journal of Reine Angewandte Mathematik, Vol. 92, No. 1, 1882, p. 56171.

11. Y. Jiang and M. Liu, "Granular Elasticity without the Coulomb Condition," Physical Review Letters, Vol. 91, No. 14, 2003, pp. 144301: 1-4.

12. Y. Jiang and M. Liu, "A Brief Review of 'Granular Elasticity'," European Physical Journal E, Vol. 22, No. 3, 2007, pp. 255-260.

13. Y. Jiang and M. Liu, "Granular Solid Hydrodynamics," Granular Matter, Vol. 11, No. 3, 2009, pp. 139-156.

14. D. Serero, G. Rydellet, E. Clément and D. Levine, "Stress Response Function of a Granular Layer: Quantitative Comparison between Experiments and Isotropic Elasticity," European Physical Journal E, Vol. 6, No. 2, 2001, pp. 169-179.

15. W. G. Ellenbroek, M. van Hecke and W. van Saarloos, "Jammed Frictionless Disks: Connecting Global and Local Response," Physical Review E, Vol. 80, No. 6, 2009, pp. 061307: 1-18.

16. C. Goldenberg and I. Goldhirsch, "Force Chains, Microelasticity and Macroelasticity," Physical Review Letters, Vol. 89, No. 8, 2002, pp. 0843021: 4.

17. I. Goldhirsch and C. Goldenberg, "On the Microscopic Foundations of Elasticity," European Physical Journal E, Vol. 9, No. 3, 2002, pp. 245-251.

18. C. Goldenberg and I. Goldhirsch, "Small and Large Scale Granular Statics," Granular Matter, Vol. 6, No. 2-3, 2004, pp. 87-96.

19. C. Goldenberg and I. Goldhirsch, "Friction Enhances Elasticity in Granular Solids," Nature, Vol. 435, No. 7039, 2005, pp. 188-191.

20. G. R. Liu and M. B. Liu, "Smoothed Particle Hydrodynamics: A Meshfree Particle Method," World Scientific, Singapore City, 2003.

21. T. S. Majmudar, M. Sperl, S. Luding and R. P. Behringer, "Jamming Transition in Granular Systems," Physical Review Letters, Vol. 98, No. 5, 2007, pp. 058001: 1-4.

22. D. O. Krimer, M. Pfitzner, K. Bräuer, Y. Jiang and M. Liu, "Granular Elasticity: General Considerations and the Stress Dip in Sand Piles," Physical Review E, Vol. 74, No. 6, 2006, pp. 061310: 1-10.

23. K. Bräuer, M. Pfitzner, O. Krimer, Y. Jiang and M. Liu, "Granular Elasticity: Stress Distributions in Silos and Under Point Loads," Physical Review E, Vol. 74, No. 6, 2006, pp. 061311: 1-10.

24. J. D. Goddard, "Nonlinear Elasticity and Pressure-Dependent Wave Speeds in Granular Media," Proceedings of the Royal Society of London A, Vol. 430, No. 1878, 1990, pp. 105-131.

25. P.-G. de Gennes, "Static Compression of a Granular Medium: The 'Soft Shell' Model," Europhysics Letters, Vol. 35, No. 2, 1996, pp. 145-149.

26. K. L. Johnson, K. Kendall and A. D. Roberts, "Surface Energy and the Contact of Elastic Solids," Proceedings of the Royal Society of London A, Vol. 324, No. 1558, 1971, pp. 301-313.

CHAPTER 8

First Evidence of Surface SH-Wave Propagation in Cubic Piezomagnetics

Aleksey Zakharenko

International Institute of Zakharenko Waves (IIZWs),
Krasnoyarsk, Russia.

ABSTRACT

This theoretical work provides with results of characteristics calculation of the ultrasonic surface Zakharenko waves (USZWs) existing in piezomagnetic cubic monocrystals of class m3m that can be readily used for non-destructive testing. The piezomagnetic waves propagate in direction [**101**] corresponding to relatively easy magnetization for the following piezomagnetics: Galfenol, Terfenol-D, and $CoFe_2O_4$ with cubic structures. The phase velocities of the USZW-waves and the coefficient of magnetomechanical coupling (CMMC) K^2 were calculated for the crystals. It was found that the coefficient K^2 for piezomagnetics with $K_m^2 > 1/3$ and $K_m^2 >> 1/3$ is about 8% to 9%, where $K^2 = 2 (V_{USZW,o} - V_{USZW,c})/V_{USZW,o}$ and $K_m^2 = h^2/(C\mu)$. Knowledge of piezomagnetic properties of cubic crystals makes possible the use of them in new products utilizing the phenomenon called the magnetoelectric effect. Also, this study is useful for possible application of cubic piezomagnetics in composite structures consisting of piezoelectric and (or) piezomagnetic materials and in the microwave technology. This broadens choice of possible piezomagnetic materials for utilization in various technical devices.

KEYWORDS

piezomagnetic cubic monocrystals, Galfenol and Terfenol-D, ultrasonic surface Zakharenko waves

1. INTRODUCTION

Ferroic materials can display a spontaneous magnetization (ferromagnetic), polarization (ferroelectric) and strain (ferroelastic, shape memory alloy). Materials possessing two "ferro" properties simultaneously are called "biferroics". There are three kinds of biferroic materials: electroelastic, magnetoelastic, and magnetoelectric materials. The magnetoelectric materials (composites) can simultaneously possess ferromagnetic and ferroelectric properties. Composite materials or structures consisting of piezoelectric and piezomagnetic phases are able to facilitate the conversion of energy between electric and magnetic fields. Such phenomenon is called magnetoelectric (ME) effect. The ME effect of the composites was first reported by van Suchtelen [1]. The ME effect was then studied by van den Boomgaard [2], van Run et al. [3], and van den Boomgaard et al. [4] for $BaTiO_3$/ $CoFe_2O_4$ composites. Possible applications of ME materials include magnetic-electric energy converting components; solid state non-volatile memory, multi-state memory, which can find application in quantum computing area; and electrical/optical polarization components, which can find applications in communication, light computing, and solid state memories based on spintronics [5,6]. Also, this research topic can be found in a recent review paper [7] by Fiebig. Note that the ME effect was originally predicted by P. Curie [8] in 1894.

Surface acoustic wave (SAW) devices are widely used in numerous branches of science and technology and their investigation especially in the case of interconnected physical fields is an important and actively developing branch of research and applications. About 120 years ago the first type of SAW was described by Lord Rayleigh [9] in connection with the problem of earthquakes. Bleustein [10] and Gulyaev [11] theoretically predicted that a pure shear-horizontal (SH) surface wave can be guided by the free surface of a piezoelectric half-space. Later, Maerfeld and Tournois [12] investigated SH acoustic waves guided by the interface of two half-spaces, and Danicki [13] described the acoustic waves which can be guided by an embedded conducting plane in the electro-elastic materials of class 6 mm. It is thought that the simplest type of SH-SAW represents Love waves [14] in layered systems, consisting of an isotropic layer and an isotropic substrate. Recent developments in physics and technology made possible to construct new magneto-electro-elastic materials which demonstrate interconnection between magnetic, electricand elastic fields [15,16]. When the electric field was connected with the elastic one (piezoelectric materials) it brought up new and unexpected possibilities for science and technology. Connecting the magnetic field with the electric and elastic ones in magneto-electro-elastic materials suggests a range of new possibilities. Alshits et al. [17] conducted a qualitative investigation on the existence of surface waves in half-infinite anisotropic elastic media with piezoelectric, piezomagnetic, and ME effects. Soh and Liu [18] gave the existence conditions of interfacial SH-waves in a piezoelectric/piezomagnetic bi-material. Although much attention has been recently concentrated on magneto-electro-elastic materials and several dynamic problems have been solved by Hu and Li [19], Li [20], Chen, et al., [21], etc., the investigation of SAW propagation in transversely isotropic magneto-electro-elastic materials [22-25] of class 6 mm is currently an open actual subject. In addition to the transversely isotropic materials there are piezoelectric and piezomagnetic crystals with cubic symmetries, in which the surface BG-waves [10,11] cannot exist, according to [26] by Gulyaev and Hickernell. That is also true for the interfacial MTwaves [12]. However, new SH-SAWs called the ultrasonic surface Zakharenko waves (USZWs) were recently discovered in [27], which can propagate in cubic piezoelectrics. That is also true for cubic piezomagnetics. The purpose of this paper is to show the first evidence of the USZW existence in cubic piezomagnetics. It is possible to introduce piezomagnetic cubic crystals which are today highly-called for different technical devices.

The Cobalt Ferrite $CoFe_2O_4$ ($CoO \cdot Fe_2O_3$ or CFO) has the spinel ferrite structure with a stoichiometry denoted by AB_2O_4, where A is a divalent transition metal ion or a mixture of a trivalent and a monovalent metal ion. A spinel (see figure 1 and [28]) is a cubic structure in which O^{2-} ions form an fcc-lattice. It requires eight Formula equivalent AB_2O_4 to form a repeating unit cell which contains 32 O^{2-} ions, 64 tetragonal sites (A sites) and 32 octahedral sites (B sites). However, only 8 tetragonal and 16 octahedral sites are occupied so that the ratio of A-atoms/B-atoms is 1:2. CFO is known to have highly anisotropic magnetic properties and has been widely used in magnetic media and microwave device applications.

In 1971, the U.S. Navy developed an alloy of rare metals Tb and Dy with iron to create Terfenol-D ($Tb_xDy_{1-x}Fe_y$). The name "Terfenol-D" comes from a combination of TER for terbium, FE for iron, NOL for Naval Ordnance Laboratory, and D for dysprosium. Recently an iron/gallium alloy termed Galfenol has also been discovered at the Naval Surface Warfare Center (former NOL), where the magnetostrictive material Terfenol-D was discovered. It was found that magnetostriction in iron/gallium alloys peaks at a volume fraction of ~17% gallium. Unlike Terfenol-D representing a giant magnetostrictive material useable in practical operating conditions, FeGa-alloys are tough and not toxic, and can be machined and used without any special handling in devices where Terfenol-D may fracture. The cubic structure of Galfenol is shown in figure 2. Terfenol-D has been commercially available since the late 1980's and has since grown into an international industry. The Terfenol-D rods can be readily manufactured using the freestand-zone-melt process resulting in a cubic crystal structure [29] featuring a large magnetic anisotropy. Also, FeGa-alloys appear promising in filling the role as a mechanically robust material with substantial magnetostrictive capability [30]. Additional benefits of FeGa-alloys include substantially lower material costs than rare-earth based magnetostrictive alternatives. For example, the Terfenol-D ($Tb_{27}Dy_{73}Fe_{195}$) raw material cost is approximately \$ 0.50/g and current production techniques utilize carefully controlled crystal growth processes [31]. For comparison, the raw material cost for $Fe_{81}Ga_{19}$ is approximately \$ 0.08/g [31]. Indeed, Terfenol-D can be used in magnetic transducers. It is wellknown that magnetic transducers are increasingly considered as actuators and sensors for numerous aerospace, aeronautic, automotive, industrial, and biomedical applications, for example see [32-34]. Recent progress in studying the cubic structure piezomagnetic crystals Galfenol and Terfenol-D of class m3m for utilization in transducers and sensors can be additionally found in [35-38]. Section 2 theoretically describes SAW propagation in piezomagnetic cubic crystals.

2. THEORY

The rectangular coordinate system (x_1, x_2, x_3) is shown in figure 3. The suitable crystal cut and propagation direction was obtained by

45^0-rotation around the x_2-axis directed along the lowest odd-order symmetry axis of a piezomagnetic cubic crystal. Therefore, the piezomagnetic waves with the anti-plane polarization propagate along the crystal surface in direction [101]. It is noted that the x_2-axis is perpendicular to the sagittal plane. Theoretical description of wave propagation in cubic piezomagnetics is similar to that for cubic piezoelectrics, for example see [27,39,40]. The constitutive equations for a piezomagnetic material can be expressed in terms of the strains τ and the magnetic field H. Strains are related to mechanical displacements: $\tau_{ij} = (\partial U_i/\partial x_j + \partial U_j/\partial x_i)/2$ [41]. The governing mechanical equilibrium is $\partial \sigma_{ij}/\partial x_j = 0$ and the governing magnetostatic equilibrium is $\partial B_i/\partial x_i = 0$ where σ_{ij} and B_i are the stress tensor and magnetic flux, respectively. A piezomagnetic medium possesses the elastic C_{ijkl} and piezomagnetic h_{ijk} coefficients as well as the magnetic permeability coefficients μ_{ij} and the medium density r.

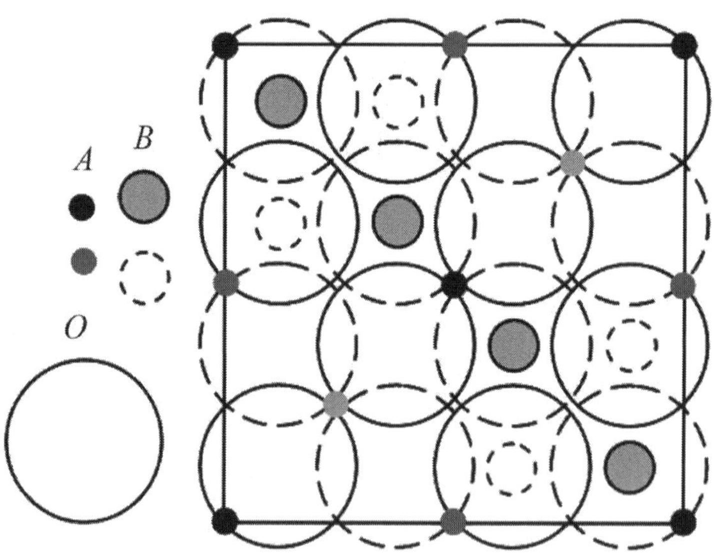

Figure 1. The normal spinel structure of Cobalt Ferrite $CoFe_2O_4(CoO{\cdot}Fe_2O_3)$ with a stoichiometry denoted by AB_2O_4. Only the lower half of the unit cell is shown for projection on (001)

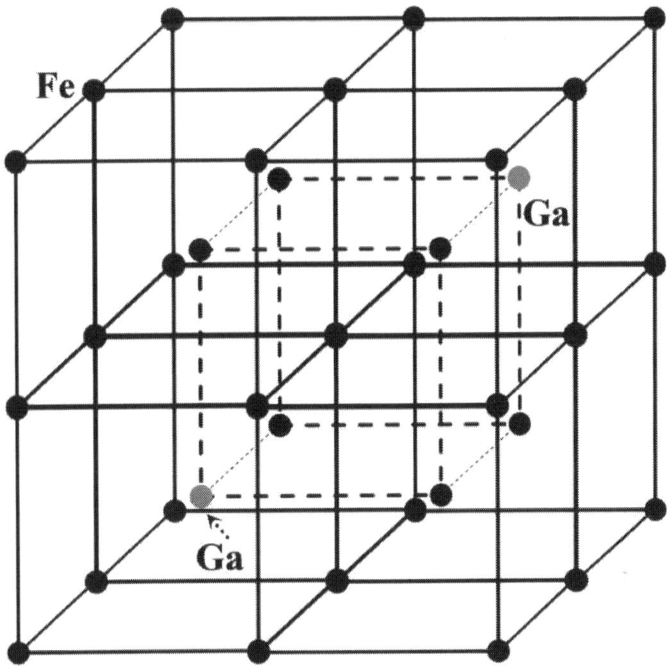

Figure 2. The bcc-structure of the Fe-17Ga alloy (α-Fe)

It is necessary to write governing equations for the linear case. Constitutive relations are written as follows:

$$\sigma_{ij} = C^{H}_{ijkl}\tau_{kl} - h_{ijm}H_{m} \tag{1}$$

$$B_{m} = h_{mij}\tau_{ij} + \mu^{\tau}_{mn}H_{n} \tag{2}$$

in which σ_{ij} and τ_{ij} are the stress and strain tensors, respectively; B_m and H_m are components of the magnetic induction (i.e. magnetic flux) and magnetic field: $H_m = -\partial\psi/\partial x_m$, where ψ is the magnetic potential. The indices i, j, k, l, m, and n run from 1 to 3. According to Voigt's usual notation, C_{ijkl}, h_{ijm}, and μ_{mn} can be written as 6 × 6, 3 × 6, and 3 × 3 matrices standing for the elastic, piezomagnetic, and magnetic tensors, respectively. They are thermodynamically defined as follows:

$$C_{ijkl}^{H} = \left(\frac{\partial \sigma_{ij}}{\partial \tau_{kl}} \right)_{H} \tag{3}$$

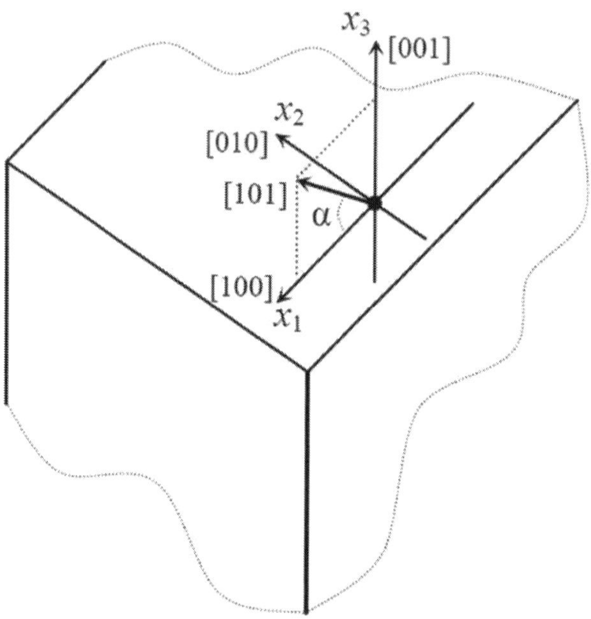

Figure 3. The rectangular coordinate system for a piezomagnetic cubic monocrystal, α = 45°

$$h_{mij} = h_{ijm} = -\left(\frac{\partial \sigma_{ij}}{\partial H_m} \right)_{\tau} = \left(\frac{\partial B_m}{\partial \tau_{ij}} \right)_{H} \tag{4}$$

$$\mu_{mn}^{\tau} = \left(\frac{\partial B_m}{\partial H_n} \right)_{\tau} \tag{5}$$

where the elastic constants C_{ijkl} are defined at constant magnetic field and the magnetic constants μ_{mn} are defined at constant strain.

In the quasi-static approximation, the equations of motion of an elastic medium and magnetostatics are

$$\frac{\partial \sigma_{ij}}{\partial x_j} = \rho \frac{\partial^2 U_i}{\partial t^2} \text{ and } \frac{\partial B_i}{\partial x_j} = 0 \quad (6)$$

In Equation (6), U_i denote the mechanical displacement components; t is time. Using Equations (1), (2), (6) and $\psi = U_4$, one can write the coupled equations of motion for a piezomagnetic medium in the following form:

$$\rho \frac{\partial^2 U_i}{\partial t^2} = C_{ijkl} \frac{\partial^2 U_l}{\partial x_j \partial x_k} + h_{kij} \frac{\partial^2 \psi}{\partial x_j \partial x_k}$$
$$0 = h_{ijk} \frac{\partial^2 U_k}{\partial x_i \partial x_j} - \mu_{ij} \frac{\partial^2 \psi}{\partial x_i \partial x_j} \quad (7)$$

Solutions of homogeneous partial differential Equations (7) of the second order are found in the following plane wave form: $U_i = U_i^0 \exp[j(\mathbf{K}_s\mathbf{R} - \omega t)]$ and $\psi = \psi^0 \exp[j(\mathbf{K}_s\mathbf{R} - \omega t)]$ where the index i runs from 1 to 3. U_i^0 and ψ are initial amplitudes; $j = (-1)^{1/2}$. $k_s r$ denotes the scalar multiplication of two vectors and ω is the angular frequency. $\{k_1, k_2, k_3\} = k\{n_1, n_2, n_3\}$ are the components of the wavevector k_s, $\{x_1, x_2, x_3\}$ are the components of the real space vector r, and $\{n_1, n_2, n_3\}$ are the directional cosines.

The coupled equations of motion can be readily written in the following simplified form, leaving only equations for waves with polarization perpendicular to the sagittal plane as well as non-zero components of the material tensors for the studied direction [101] of wave propagation:

$$\rho\frac{\partial^2 U_2}{\partial t^2} = C_{44}\left(\frac{\partial^2 U_2}{\partial x_1^2} + \frac{\partial^2 U_2}{\partial x_3^2}\right) - h_{16}\frac{\partial^2 \psi}{\partial x_1^2} + h_{34}\frac{\partial^2 \psi}{\partial x_3^2}$$

$$0 = -h_{16}\frac{\partial^2 U_2}{\partial x_1^2} + h_{34}\frac{\partial^2 U_2}{\partial x_3^2} - \mu_{11}\left(\frac{\partial^2 \psi}{\partial x_1^2} + \frac{\partial^2 \psi}{\partial x_3^2}\right) \quad (8)$$

In Equation (8), the mechanical displacement component U_2 is directed along the x_2-axis, see figure 3:

$$U_{2,4} = U_{2,4}^0 \exp\left[jk\left(n_1 x_1 + n_3 x_3 - V_{ph} t\right)\right] \quad (9)$$

where the phase velocity is defined as $v_{ph} = w/k$ (k is the wavenumber in direction of wave propagation). For piezomagnetic cubic crystals, it is possible to write $C_{44} = C_{66} = C$ and $\mu_{11} = \mu_{33} = \mu$. It is noted that many propagation directions perpendicular to direction [010] can exist. It is also possible to cut a cubic crystal in order to study wave propagation in direction [101] with $h_{14} = h_{36} = 0$ and the non-zero piezomagnetic constants $\{-h_{16}, h_{34}\}$ that is shown in Equation (8). The dependence of the normalized piezomagnetic constants on the cubic crystal cuts and propagation directions is shown in figure 4.

Substituting the mechanical displacement U_2 and magnetic potential $\psi = U_4$ of Equation (9) into Equation (8), the equations of motion can readily be written in the wellknown tensor form, using corresponding GL-components in the Green-Christoffel equation, $(GL_{rw} - \delta_{rw}\rho V_{ph})\, U_r = 0$

[39,41]: $GL_{22} = C(1+n_3^2)$, $GL_{24} = GL_{42} = -h(1-n_3^2)$ and $GL_{44} = -\mu(1+n_3^2)$ with $n_3 = k_3/k$. In the GL-equation, r and w run from 1 to 4, δ_{rw} is the Kronecker delta for $r < 4$ and $w < 4$, $\delta_{44} = 0$, $U_r = \{U_1, U_2, U_3, \psi\}$. Therefore, the following system of two homogeneous equations for pure SH-waves can be written as:

$$\begin{pmatrix} \mathrm{GL}_{22} - C(V_{ph}/V_{t4})^2 & \mathrm{GL}_{24} \\ \mathrm{GL}_{42} & \mathrm{GL}_{44} \end{pmatrix} \begin{pmatrix} U_2^0 \\ \psi^0 \end{pmatrix} = 0 \tag{10}$$

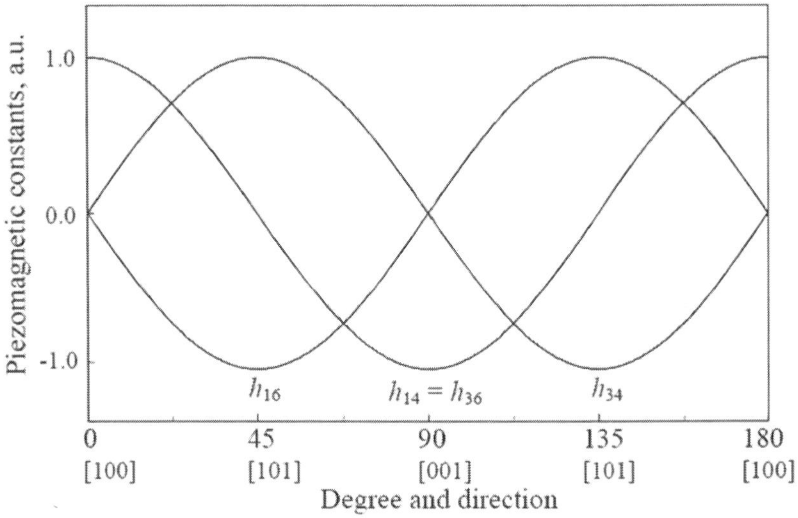

Figure 4. The dependence of the normalized piezomagnetic constants $h_{14} = h_{36}$, h_{16}, and h_{34} on the propagation directions with the Euler angles {0°, θ, 0°} for piezomagnetic cubic crystals, where there is rotation around the x_2-axis

In Equation (10), the directional cosines are defined as follows: $n_1 \equiv 1$, $n_2 \equiv 0$ and $n_3 = n_3$. The velocity V_{t4} is also defined as follows: $V_{t4} = (C_{44}/\rho)^{1/2}$. Setting the matrix determinant equal to zero in Equation (10), the suitable phase velocity V_{ph} satisfying boundary conditions discussed in the following section and four polynomial roots $n_3^{(p)}(V_{ph})$, as well as the functions $U_2^0(V_{ph})$ and $\varphi^0(V_{ph})$, can then be found. For example, the functions can be taken in the following form: $\psi^0 = \mathrm{GL}_{42}$ and $U_2^0 = -\mathrm{GL}_{44}$. It is noted that for the piezomagnetic surface Bleustein-Gulyaev waves, the GL_{24} and GL_{42} components are as follows: $\mathrm{GL}_{24} = \mathrm{GL}_{42} = h(1+n_3^2)$.

Substituting $m = 1 + n_3^2$ and expanding the matrix determinant in Equation (10), the following polynomial can be obtained from Equation (10):

$$\left(1+K_m^2\right)m_3^2 - Dm_3 + 4K_m^2 = 0 \quad \text{with} \quad D = \left(\frac{V_{ph}}{V_{t4}}\right)^2 + 4K_m^2 \qquad (11)$$

of which two roots can be found as follows

$$m_3^{(1,2)} = \frac{D \pm \sqrt{D^2 - 16K_m^2\left(1+K_m^2\right)}}{2\left(1+K_m^2\right)} \qquad (12)$$

giving four polynomial roots of Equation (10)

$$n_3^{(1,2,3,4)} = \pm\sqrt{-1+m_3^{(1,2)}} \qquad (13)$$

The polynomial roots $n_3^{(1,2,3,4)}$ actually represent the eigenvalues. For each eigenvalue $n_3^{(n)}$, the corresponding eigenvector can be found in the following form: $(U_2^{(n)}, \psi^{(n)})$ where the index n runs from 1 to 4. The coefficient K_m^2 in Equations (11) and (12) is the static coefficient of the magnetomechanical coupling (CMMC):

$$K_m^2 = h_{16}^2 / \left(C_{44}\mu_{11}\right) \qquad (14)$$

It is also noted that the speed V_{tm} of the bulk SH-wave is given by the following Formula:

$$V_{tm} = V_{t4}\left(1+K_m^2\right)^{1/2} \qquad (15)$$

Analyzing the roots for propagation direction [101] in Equations (12) and (13), it can be found that all complex roots will be calculated when the expression under the square root in Equation (12) is negative. That fulfills for velocities V_{ph} being lower than some velocity V_{Km} obtained solving the following Equation from (12):

$$D^2 - 16K_m^2\left(1 + K_m^2\right) = 0 \tag{16}$$

and defined by the following formula:

$$V_{Km} = A_{Km}V_{t4} \text{ with } A_{Km} = 2\sqrt{K_m\sqrt{1+K_m^2} - K_m^2} \tag{17}$$

It is clearly seen in Equation (17) that the factor A_{Km} shown in figure 5 is a function of the CMMC K_m^2. The other function f $(K_m^2) = (1 + K_m^2)^{1/2} = V_{tm}/V_{t4}$ from Equation (15) is also shown in figure 5. The function A_{km} ($K_m^2 = K_0^2 = 1/3$) approaches the function f $(K_m^2) = (1 + K_m^2)^{1/2}$ giving the following equality: $V_{Km} = V_{tm}$ (see figure 5). It is noted that only complex polynomial roots can exist when $V_{ph} < V_{Km}$. The CMMC $K_0^2 = 1/3$ is readily found by substituting the velocity V_{tm} from Equation (15) instead of the phase velocity V_{ph} in Equation (16). Note that when $K_m^2 < K_0^2$ there are all imaginary roots for $V_{ph} > V_{Km}$, but a big $K_m^2 > K_0^2$ gives real roots when $V_{ph} > V_{Km}$. It is also noted that only complex or imaginary roots with negative imaginary parts are chosen in order to cope with wave damping towards the depth of a crystal corresponding to negative values of the x_3-axis shown in figure 3.

Table 1 lists experimental data for Galfenol (83% of Fe and 17% of Ga) obtained in [42,43]. It is clearly seen that the static CMMC K_m^2 for cubic Galfenol can be as high as~0.42 that is larger than 1/3, see also table 2. The Galfenol density ρ was taken from [31], which can depend on material synthesis. For example, [31] gives a Galfenol density of ~5900 kg/m^3 for a typical sintered disk that is only 74% of maximum possible density. It is noted that the material density of Fe is ~7848 kg/m^3 (see [44]). According to [31,42] as well as many Galfenol Workshops at the University of Maryland, the shear elastic constant c_{44} = c for Galfenol relatively slightly depends on both temperature and concentration of ga in fe.

The elastic constant c can have values from ~120 GPa to ~135 gPa. According to [31], the static CMMC K_m^2 for cubic Terfenol-D can be usually about two times larger than that for Galfenol under the same conditions. However, it is interesting to compare some characteristics of Galfenol and Terfenol-D. Therefore, Terfenol-D with a very big K_m^2 ~1.2 was chosen from [45] for numerical calculations suggesting that Terfenol-d has the cubic symmetry. The piezomagnetics $CoFe_2O_4$ [16,46] is also listed in table 2 for comparison, suggesting the cubic symmetry for the crystal.

Since the experimental data listed in **Table 1** is given for different experimental conditions, it is possible to calculate the single piezomagnetic constant $h = h_{14}$ for cubic Galfenol, using known values of K_m and C, as well as the measured values of magnetic permeability μ given in table 1. The strain τ and magnetic flux B can be also written based on the following linear constitutive piezomagnetic equations:

$$\tau = S^H \sigma + qH \tag{18}$$

$$B = q^* \sigma + \mu^\sigma H \tag{19}$$

where s^H is the compliance at constant magnetic field H and μ^σ is the magnetic permeability at a constant stress σ. Note that Equation (19) incorporates the direct magnetostrictive effect in which magnetic flux is generated by stresses in the piezomagnetic material. The piezomagnetic constants q and q^* are thermodynamically defined as follows, assuming $q = q^*$:

$$q = (\partial\tau/\partial H)_\sigma \text{ and } q^* = (\partial B/\partial\sigma)_H \tag{20}$$

The piezomagnetic coefficient q symbolizes the axial magnetostriction coefficient and represents the change in strain per change in magnetic field at a constant stress. The parameter q* represents the change in magnetic induction per change in stress at a constant magnetic field.

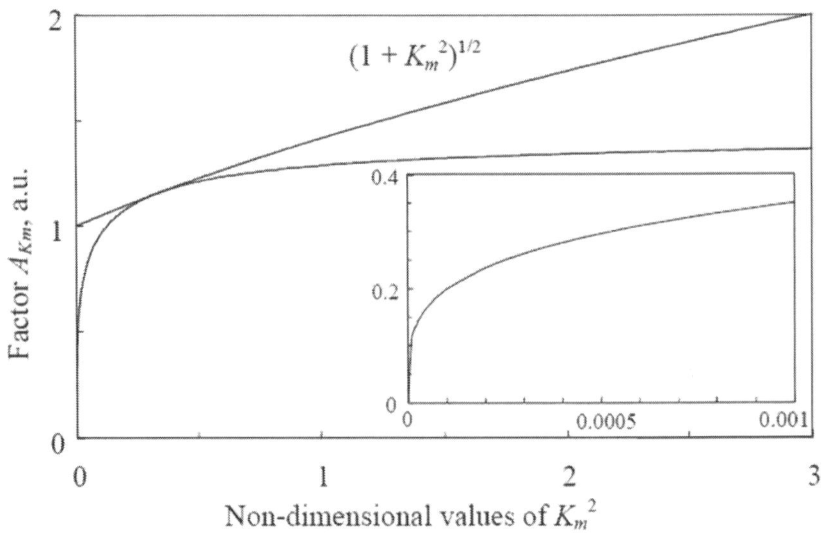

Figure 5. The dependence of both the factor A_{Km} and the function $(1 + K_m^2)^{1/2}$ on the values of K_m^2 from 0 up to 3. The insertion shows the $A_{Km}(K_m^2)$ for small values of the K_m^2

The short circuit elastic modulus E^B represents the stiffest-material condition, which occurs when all available magnetic energy has been transduced into elastic potential energy. When energy is transferred from the elastic to the magnetic regime the effective modulus decreases to the value of E^H. It has been shown by invoking linear transduction considerations that these two elastic moduli are related by the magnetomechanical coupling coefficient K_m^2 ($0 \leq K_m^2 \leq 1$) as follows [47]:

$$E^H = E^B\left(1 - K_m^2\right) \tag{21}$$

Table 1. The material characteristics following [42,43] for Galfenol $Fe_{0.83}Ga_{0.17}$ (m3m); $E^B = 7.7 \times 10^{10}$ [N/m^2]. The magnetic permeability coefficient of a vacuum is $\mu_0 = 4\pi \times 10^{-7}$ [H/m]~12.566371 × 10^{-7} [(V × s)/(A × m)]. The static coefficient of magnetomechanical coupling (CMMC) K_m in the last column was also announced in [42,43]

Stress Ξ, 10^6 [N/m^2]	q, 10^{-9} [m/A]	μ^σ/μ_0,	K_m
16.0	58	285	0.65
27.5	34	174	0.54
50.5	22	108	0.47
85.0	16	70	0.42

In a similar manner, the intrinsic or uncoupled magnetic permeability μ^σ of Equation (19) can be reduced to a value corresponding to the constant strain permeability μ^ε because of the energy conversion from the magnetic to the elastic regime. The permeabilities are related by:

$$\mu^\varepsilon = \mu^\sigma\left(1 - K_m^2\right) \tag{22}$$

It is emphasized that expressions (21) and (22) are only valid for linear systems and significant corrections are necessary in order to extend these formulae to the full nonlinear regime. However, they are highly useful to illustrate the exchange mechanisms taking place in magnetostrictive transduction.

The static CMMC K_m^2 can be also written as follows:

$$K_m^2 = q^2 / \left(S^H \mu^\sigma\right) \tag{23}$$

Table 2. The elastic C_{44} and piezomagnetic h_{14} constants as well as the material density ρ and magnetic permeability coefficient μ_{11} for the piezomagnetic cubic crystals: Galfenol [42,43], $CoFe_2O_4$ [16,46], and Terfenol-D [45]. The coefficient of magnetomechanical coupling (CMMC) K_m^2 in the last column can be calculated with Formula (14)

Material (symmetry group)	Density, ρ [kg/m^3]	C_{44}, 10^{10} [N/m^2]	h_{14} [N/(A×m)]	$\mu_{11} = \mu'$, 10^{-6} [N/A^2]	K_m^2
Galfenol (m3m)	7973	12.70	3331.34	206.83	0.4225
Galfenol (m3m)	7973	12.70	2395.05	154.90	0.2916
Galfenol (m3m)	7973	12.70	1722.32	105.74	0.2209
Galfenol (m3m)	7973	12.70	1273.98	72.45	0.1764
$CoFe_2O_4$ (m3m)	5300	4.53	550.0	157.00	0.0425
Terfenol-D (m3m)	9250	0.60	167.7	3.97	1.1807

Therefore, the single non-zero piezomagnetic constant h can be calculated with the following formula, using Equations (14), (22), and (23):

$$h = K_m \sqrt{C\mu^\sigma \left(1 - K_m^2\right)} \tag{24}$$

Table 2 lists the piezomagnetic constant h for cubic Galfenol corresponding to the experimentally measured values of the magnetic permeability μ^σ at constant stress σ.

For the free space, Laplace's equation of type $D\psi = 0$ is written in the following form: $(k_1^2 + k_3^2)\,\psi_0 = 0$ where μ_0 is the magnetic constant for the free space. The magnetic potential for the free space can be written as follows: $\psi_0 = F^{(0)}\exp(-k_1x_3)\exp[j(k_1x_1 - wt)]$. The potential ψ_0 must decrease according to any increase in the coordinate x_3 (see figure 3). It is now possible to use the material constants in mechanical and magnetic boundary conditions.

3. BOUNDARY CONDITIONS FOR SURFACE SH-WAVES

The mechanical boundary conditions of a magnetoelastic material which occupies the half-space $x_3 < 0$ must be satisfied together with the magnetic conditions of a magnetically closed ($B_3 = 0$) or magnetically open ($\psi = 0$) surface. The realization of each of these boundary conditions is described in [17]. There is the single mechanical boundary condition for the normal component of the stress tensor: $\sigma_{32} = 0$ at $x_3 = 0$, where,

$$\sigma_{32} = \sum_{p=1,2} F^{(p)}\left[Ck_3^{(p)}U_2^{(p)} + hk_3^{(p)}U_4^{(p)}\right] \tag{25}$$

Also, there are the magnetic boundary conditions: continuity of the normal component B_3 of the magnetic flux at $x_3 = 0$ being the interface between a vacuum (B_3^f) and the crystal surface ($B_3 = B_3^f$) where,

$$B_3 = \sum_{p=1,2} F^{(p)}\left[hk_3^{(p)}U_2^{(p)} - \mu k_3^{(p)}U_4^{(p)}\right]$$

$$B_3^f = -F^{(0)}\psi_0^f \, jk_1\mu_0 \tag{26}$$

and continuity of the magnetic potential $U_4 = \psi$ at $x_3 = 0$ ($\psi = \psi^f$) where

$$\psi = \sum_{p=1,2} F^{(p)}\psi^{(p)} \quad \text{and} \quad \psi^f = F^{(0)}\psi_0^f \tag{27}$$

Therefore, using Equations (25)-(27), two homogeneous equations for the case of a magnetically closed surface can be readily written as follows:

$$\begin{pmatrix} Ck_3^{(1)}U_2^{(1)} + hk_3^{(1)}U_4^{(1)} & Ck_3^{(2)}U_2^{(2)} + hk_3^{(2)}U_4^{(2)} \\ j\mu_0 k_1 U_4^{(1)} & j\mu_0 k_1 U_4^{(2)} \end{pmatrix}\begin{pmatrix} F^{(1)} \\ F^{(2)} \end{pmatrix} = 0 \tag{28}$$

For a magnetically open surface, the following homogeneous equations can also be written in the corresponding form:

$$\begin{pmatrix} Ck_3^{(1)}U_2^{(1)} + hk_3^{(1)}U_4^{(1)} & Ck_3^{(2)}U_2^{(2)} + hk_3^{(2)}U_4^{(2)} \\ hk_3^{(1)}U_2^{(1)} - (\mu k_3^{(1)} - j\mu_0 k_1)U_4^{(1)} & hk_3^{(2)}U_2^{(2)} - (\mu k_3^{(2)} - j\mu_0 k_1)U_4^{(2)} \end{pmatrix}\begin{pmatrix} F^{(1)} \\ F^{(2)} \end{pmatrix} = 0 \tag{29}$$

Therefore, the corresponding boundary-condition determinants (BCDs) of matrices in Equations (28) and (29) for both cases represent complex numbers.

The complete mechanical displacement U_2^Σ and magnetic potential $\psi^\Sigma = U_4^\Sigma$ can be written in the plane wave form as follows:

$$U_{2,4}^{\Sigma} = \sum_{p=1,2} F^{(p)} U_{2,4}^{0(p)} \exp\left[jk\left(n_1 x_1 + n_3^{(p)} x_3 - V_{ph} t \right) \right] \tag{30}$$

The corresponding weight functions $F^{(1)}$ and $F^{(2)}$ are found from Equations (28) and (29), which can give the same eigenvectors ($U_2^{0(1)}$, $\psi^{0(1)}$) and ($U_2^{0(2)}$, $\psi^{0(2)}$) for two equal eigenvalues $n_3^{(1)}$ = $n_3^{(2)}$, and hence $F^{(1)} = -F^{(2)}$. It is obvious that for this case the weight factors $F^{(1)} = -F^{(2)}$ will zero the complete mechanical displacement U_2^{Σ} and magnetic potential ψ^{Σ} in Equations (28) and (29) giving "latent" characteristics in Equation (30). On the other hand, unequal eigenvalues $n_3^{(1)}$and $n_3^{(2)}$ give different eigenvectors ($U_2^{0(1)}$, $\psi^{0(1)}$) and ($U_2^{0(2)}$, $\psi^{0(2)}$). It is thought that the V_{ph}-solutions with the "latent" characteristics must be experimentally verified due to the fact that some elements of crystals symmetry (screw axis or glide reflection) can be broken near the surface. The same relates to simple reflections and axis if the surface restricts the crystal in such a way that it breaks some simple elements of the crystal class group. It is also noted that V_{ph}- solutions cannot be found when the surface BleusteinGulyaev waves are studied that is an additional difference for finding the BG-waves and USZW-waves.

4. RESULTS AND DISCUSSIONS

Acoustic wave propagation along the surface of piezoelectric material in direction [101] as an instability problem can sustain the surface SH-waves in cubic piezoelectrics that was recently discovered in [27]. Concerning piezomagnetics, it is thought that piezomagnetic SH-SAW can also propagate in direction [101] in cubic piezomagnetics, for instance, crystals of class m3m. Indeed, all cubic crystals of the class possess no piezoelectricity according to [48,49]. Note that concerning transversely-isotropic materials, solutions for the surface BG-waves in piezomagnetics can be written in an identical manner to the wave solutions in piezoelectrics [50]. It is also noted that the velocity V_{BGmo} of piezomagnetic surface BG-wave propagating along the magnetically open surface of transversely-isotropic piezomagnetics (treating the simplest case) can be found with the following well-known formula:

$$V_{BGmo} = V_{tm}\left(1-\left(\frac{K_m^2}{\left(1+K_m^2\right)\left(1+\mu/\mu_0\right)}\right)^2\right)^{1/2} \tag{31}$$

The velocity V_{BGmc} of piezomagnetic BG-wave with the magnetically closed surface can be calculated with the following formula:

$$V_{BGmc} = V_{tm}\left(1-\left(\frac{K_m^2}{1+K_m^2}\right)^2\right)^{1/2} \tag{32}$$

Note that Formula (31) for the velocity V_{BGmo} is not applicable for the case of cubic piezomagnetics. For example, formula (31) gives V_{BGmo}~4760.093893 m/s for Galfenol with the largest value of $K_m^2 > 1/3$ in the first row in table 3. This value of V_{BGmo} is close to that of V_{tm}, but not to the value of V_{Km} (see table 3) representing the right result for cubic piezomagnetics with $K_m^2 > 1/3$. The same there is for Terfenol-D with K_m^2~1 in table 2, for which the incorrect velocity calculated with Formula (31) is V_{BGmo}~1179.199962 m/s.

For cubic piezomagnetics with $K_m^2 < 1/3$, Formula (31) also gives values of V_{BGmo} being closer to corresponding values of V_{tm} than true values of $V_{USZW,o}$. For example, the velocity $V_{USZW,o}$ for Galfenol in the fourth row in table 3 with the smallest value of $K_m^2 < 1/3$ (see also table 2) is $V_{USZW,o}$~4328.763803 m/s. However, Formula (31) gives V_{BGmo} ~4328.783915 m/s that is closer to V_{tm}~

4328.798060 m/s. Therefore, the differences $\Delta_{BG} = V_{BGmo} - V_{tm}$ and $\Delta_{USZW} = V_{USZW,o} - V_{tm}$ are ~0.014 m/s and ~0.034 m/s, respectively. For comparison, for the piezomagnetic $CoFe_2O_4$ with the smallest value of K_m^2 ~4% from table 2, the true velocity $V_{USZW,o}$ in table 3 is already very close to the velocity V_{BGmo} ~2985.08072 m/s, which is also very close to the velocity V_{tm}. Note that the relative magnetic permeability μ/μ_0 in Formula (31) for piezomagnetics is about one order larger than the corresponding relative dielectric constant $\varepsilon/\varepsilon_0$ for piezoelectrics for the same $K_m^2 = K_e^2$, where K_e^2 is the static coefficient of electromechanical coupling (CEMC), and ε and ε_0 are the single electric constants for a piezoelectrics and free space, respectively. That can result in mm/s-difference for piezomagnetics compared with cm/s-difference for piezoelectrics between the corresponding BG-wave velocity and bulk wave velocity V_t. That can result in significantly larger penetration depth of surface BG and USZW-waves for piezomagnetics towards negative values of the x_3-axis in figure 3. It is also noted that the velocity V_{BGmc} for a magnetically closed surface can be readily calculated with Formula (32), with which the velocity of interfacial Maerfeld-Tournois waves [12] is also calculated. The interfacial Maerfeld-Tournois waves can propagate along the interface of two transverselyisotropic half-spaces of class 6 mm. Concerning cubic piezomagnetics, it is expected that some interfacial waves can also propagate that can be further researched.

Using the velocities $V_{USZW,o}$ and $V_{USZW,c}$ for the USZWwaves, it is possible to evaluate the coefficient of magnetomechanical coupling (CMMC) K^2 for the piezomagnetic cubic crystals listed in table 3. Indeed, it is thought that it is possible to use the well-known formula that is used for piezoelectrics. Therefore, the coefficient K^2 can be evaluated with the following formula:

$$K^2 = 2\left(\frac{V_{USZW,o} - V_{USZW,c}}{V_{USZW,o}}\right) \tag{33}$$

Table 3. The velocities v_{Km}, V_{t4}, and V_{tm} (all in m/s) as well as the velocities $V_{USZW,o}$ and $V_{USZW,c}$ of the ultrasonic surface Zakharenko waves for [101] propagation direction in the piezomagnetic cubic crystals. The last column gives values of the K^2 calculated with Formula (33). All the Galfenols correspond to those listed in tables 1 and 2

Material (symmetry group)	Structure	V_{Km}	V_{t4}	V_{tm}	$V_{USZW,o}$	$V_{USZW,c}$	K^2, %
Galfenol (m3m)	bcc	4740.796256	3991.085010	4760.101550	4740.750567	4545.299858	8.24
Galfenol (m3m)	bcc	4530.187138	3991.085010	4535.795584	4535.727732	4418.693612	5.16
Galfenol (m3m)	bcc	4360.495418	3991.085010	4409.913035	4409.878451	4337.133082	3.30
Galfenol (m3m)	bcc	4217.255080	3991.085010	4328.798060	4328.763803	4279.858496	2.26
$CoFe_2O_4$ (m3m)	Spinel	2396.906402	2923.554311	2985.080877	2985.080691	2982.595557	0.17
Terfenol-D (m3m)	-	1048.738635	805.3872662	1189.319685	1047.759141	999.920958	9.13

For Galfenol in tables 2 and 3, the evaluated value of K^2 is as high as 2.26% for the smallest value of K_m^2. The value of K^2 increases with increase in K_m^2 to the biggest value of 8.25%. This value of 8.25% for the piezomagnetic Galfenol in the first row of table 3 can be compared with that for Terfenol-D in the Table: K^2(Terfenol-D)~ 9.13%. It is clearly seen in the second and third tables that a big value of K_m^2~1.2 for Terfenol-D being about three times larger than the biggest value of K_m^2 for Galfenol does not give a significant rise in the coefficient K^2. The rise is smaller than 1%. That can be explained using the following fact: the velocity $V_{USZW,o}$ for cubic piezomagnetics with a giant coefficient K_m^2 is situated slightly below the speed V_{Km}, but not slightly below the speed V_{tm}(see also figure 5). That results in lower coefficient K^2. Therefore, it can be concluded that the coefficient K^2 for cubic piezomagnetics with a giant K_m^2 is about 8% to 9% and cannot be larger than 10%. It is thought that the coefficient K^2 for the surface BG-waves in transversely-isotropic piezomagnetics with a giant K_m^2 can be larger than 10%, according to Formulae (31)–(33) and figure 5. The sample behaviors of the boundarycondition determinant (BCD3) around the phase velocity solutions for the piezomagnetic monocrystals $CoFe_2O_4$ and Galfenol are shown in figures 6 and 7, respectively. **Figure 8** shows the dependence of the V_{ph} on the coefficient K_m for the Fe-17Ga alloy. Note that any change in the material density will change both the velocities V_{t4} and $V_{tm}(V_{t4})$, but not the coefficients K_m^2 and K^2.

5. CONCLUSIONS

This paper addresses the theoretical investigations of piezomagnetic acoustic waves propagating along the surface of cubic crystals. The ultrasonic surface Zakharenko waves (USZWs) can propagate in cubic piezomagnetics when direction [101] is parallel to the crystal surface. The computer simulation was done for the widelycalled piezomagnetic crystals Galfenol, Terfenol-D, and $CoFe_2O_4$ suggesting the cubic symmetry. The cubic piezomagnetics Galfenol and Terfenol-D can possess large values of K_m^2, where K_m^2 is the static coefficient of magnetomechanical coupling. The simulations have shown that for a cubic crystal with a large $K_m^2 > 1/3$ (even $K_m^2 > 1$ for Terfenol-D) the coefficient K^2 is as big as 8%-9% and not larger than 10%. It is noted that a big $K_m^2 > 1$ does not give a significant rise to the coefficient K^2 for the case of $K_m^2 \sim 1/3$. That is due to the fact that velocity solutions for large values of $K_m^2 > 1/3$ are found just below the velocity V_{Km}, but not just below the velocity V_{tm}. This theoretical study of cubic piezomagnetics allows use of suitable surfaces of piezomagnetic cubic crystals in non-destructive testing and evaluation that was also recently done for cubic piezoelectrics.

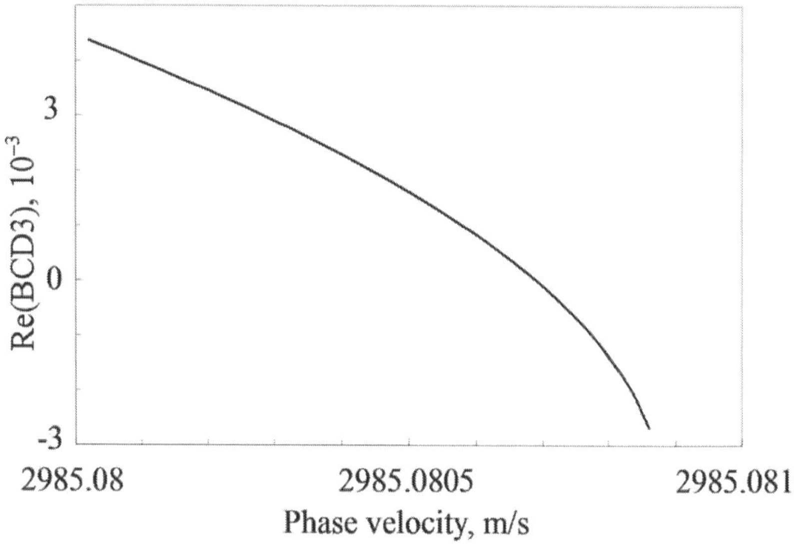

Figure 6. The behavior of the boundary-condition determinant (BCD3) for the piezomagnetic monocrystal $CoFe_2O_4$

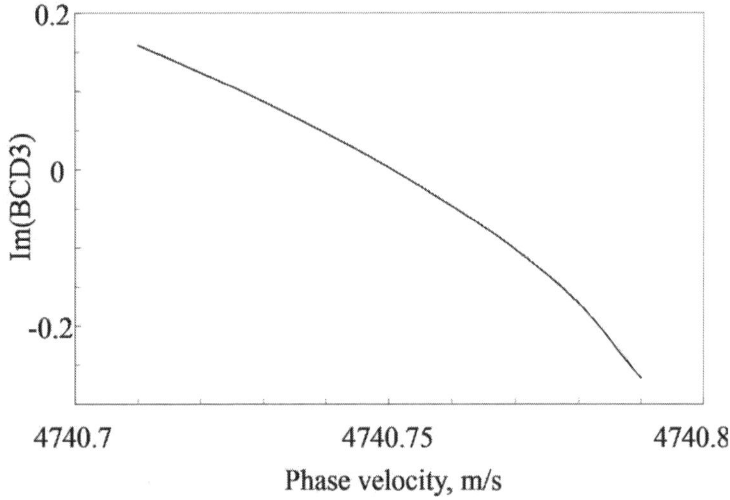

Figure 7. The behavior of the boundary-condition determinant (BCD3) for the piezomagnetic bcc-cubic monocrystal of the Fe-17Ga alloy for $K_m = 0.65$

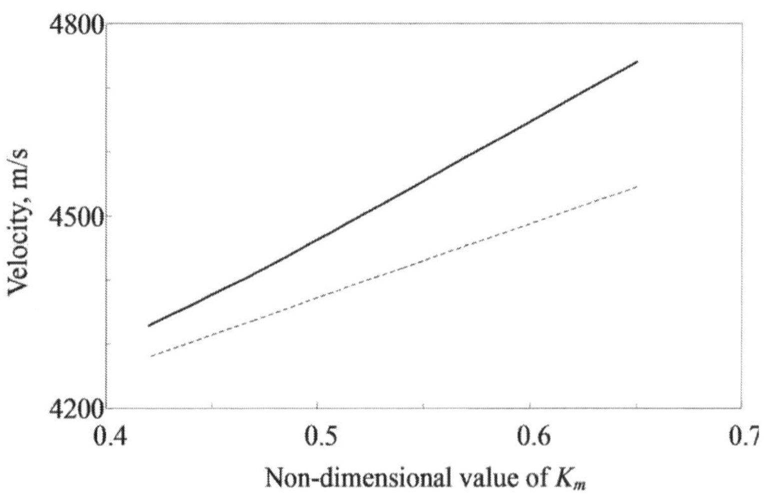

Figure 8. The dependence of the phase velocity V_{ph} on the coefficient $K_m$for the Fe-17Ga alloy, where the functions $V_{USZW,o}(k_m)$ and $V_{USZW,c}(k_m)$ are shown by solid and dashed lines, respectively

ACKNOWLEDGEMENTS

The author is grateful to all the Referees for their constructive comments.

REFERENCES

1. J. V. Suchtelen, "Product Properties: A New Application of Composite Materials," Philips Research Reports, Vol. 27, No. 1, 1972, pp. 28-37.

2. J. V. D. Boomgaard, D. R. Terrell, R. A .J. Born and H. F. J. I. Giller, "An in Situ Grown Eutectic Magnetoelectric Composite-Material: 1. Composition and Unidirectional Solidification," Journal of Materials Science, Vol. 9, No. 10, 1974, pp. 1705-1709.

3. M. J. G. V. Run, D. R. Terrell and J. H. Scholing, "An in Situ Grown Eutectic Magnetoelectric Composite-Material: 2. Physical Properties," Journal of Materials Science, Vol. 9, No. 10, 1974, pp. 1710-1714.

4. J. V. D. Boomgaard, A. M. J. G. V. Run and J. V. Suchtelen, "Piezoelectric-Piezomagnetic Composites with Magneto-electric Effect," Ferroelectrics, Vol. 14, 1976, pp. 727-728.

5. V. E. Wood and A. E. Austin, "Magnetoelectric Interaction Phenomena in Crystals," In A. J. Freeman, H. Schmid, Eds., Gordon and Breach Science Publishers, Newark, New York, 1975, pp. 181-194.

6. Y. N. Venevtsev, V. V. Gagulin and I. D. Zhitomirsky, "Material Science Aspects of Seignette-Magnetism Problem," Ferroelectrics, Vol. 73, 1987, pp. 221-248.

7. M. Fiebig, "Revival of the Magnetoelectric Effect," Journal of Physics D: Applied Physics, Vol. 38, No. 8, 2005, pp. 123-152.

8. P. Curie, "Sur la Symétrie des Phénomènes Physiques: Symétrie d'un Champ électrique et d'un Champ Magnétique," Journal de Physique, 3e Série, Vol. 3, 1894, pp. 393-415.

9. L. Rayleigh, "On Waves Propagated along the Plane Surface of Elastic Solids," Proceedings of the London Mathematical Society, Oxford, Vol. 17, 1885, pp. 4-11.

10. J. L. Bleustein, "A New Surface Wave in Piezoelectric Materials," Applied Physics Letters, Vol. 13, No. 12, 1986, pp. 412-413.

11. Y. V. Gulyaev, "Electroacoustic Surface Waves in Solids," Soviet Physics Journal of Experimental and Theoretical Physics Letters, Vol. 9, 1969, pp. 37-38.

12. C. Maerfeld and P. Tournois, "Pure Shear Elastic Surface Wave Guide by the Interface of two Semi-Infinite Media," Applied Physics Letters, Vol. 19, No. 4, 1971, pp. 117-118.

13. E. Danicki, "New Interfacial Shear Wave in Piezoelectrics," Applied Physics Letters, Vol. 64, No. 8, 1994, pp. 969-970.

14. E. H. Love, "Some Problems of Geodynamics," Cambridge University Press, London, 1911.

15. W. Nan, "Magnetoelectric Effect in Composites of Piezoelectric and Piezomagnetic Phases," Physical Review B, Vol. 50, No. 9, 1994, pp. 6082-6088.

16. S. Srinivas, J. Y. Li, Y. C. Zhou and A. K. Soh, "The Effective Magnetoelectroelastic Moduli of Matrix-Based Multiferroic Composites," Journal of Applied Physics, Vol. 99, No. 4, 2006, pp. (043905)1-7.

17. V. I. Al'shits and A. N. Darinskii and J. Lothe, "On the Existence of Surface Waves in Half-Infinite Anisotropic Elastic Media with Piezoelectric and Piezomagnetic Properties," Wave Motion, Vol. 16, No. 3, 1992, pp. 265- 283.

18. A.-K. Soh and J.-X. Liu, "Interfacial Shear Horizontal Waves in a Piezoelectric-Piezomagnetic Bi-Material," Philosophical Magazine Letters, Vol. 86, No. 1, 2006, pp. 31-35.

19. K.-Q. Hu and G.-Q. Li, "Constant Moving Crack in a Magnetoelectroelastic Material under Anti-Plane Shear Loading," International Journal of Solids and Structures, Vol. 42, No. 9-10, 2005, pp. 2823-2835.

20. X.-F. Li, "Dynamic Analysis of a Cracked MagnetoelectrOelastic Medium under Anti-Plane Mechanical and InPlane Electric and Magnetic Impacts," International Journal of Solids and Structures, Vol. 42, No. 11-12, 2005, pp. 3185-3205.

21. J. Chen, E. Pan and H. Chen, "Wave Propagation in Magneto-Electro-Elastic Multilayered Plates," International Journal of Solids and Structures, Vol. 44, No. 3-4, 2007, pp. 1073-1085.

22. K. Mori and M. Wuttig, "Magnetoelectric Coupling in Terfenol-D/polyvinylidenedifluoride Composites," Applied Physics Letters, Vol. 81, No. 1, 2002, pp. 100-101.

23. M. I. Bichurin, D. A. Filippov, V. M. Petrov, V. M. Laletsin, N. Paddubnaya and G. Srinivasan, "Resonance Magnetoelectric Effects in Layered Magnetostritive-Piezoelectric Composites," Physical Review B, Vol. 68, No. 13, 2003, pp. (132408)1-4.

24. K. Srinivas, G. Prasad, T. Bhimasankaram and S. V. Suryanarayana, "Electromechanical Coefficients of Magneto-

electric PZT-$CoFe_2O_4$ Composite," Modern Physics Letters B, Vol. 14, No. 17-18, 2000, pp. 663-674.

25. P. Yang, K. Zhao, Y. Yin, J.G. Wan and J.S. Zhu, "Magnetoelectric Effect in Magnetostrictive/Piezoelectric Laminate Composite Terfenol-D/$LiNbO_3$ [(zxtw)-129°/ 30°]," Applied Physics Letters, Vol. 88, No. 17, 2006, pp. (172903)1-3.

26. Y. V. Gulyaev and F. S. Hickernell, "Acoustoelectronics: History, Present State and New Ideas for a New Era," Acoustical Physics, Vol. 51, No. 1, 2005, pp. 81-88.

27. A. Zakharenko, "New Solutions of Shear Waves in Piezoelectric Cubic Crystals," Journal of Zhejiang University SCIENCE, Vol. 8, No. 4, 2007, pp. 669-674.

28. W. Kleber, "An Introduction to Crystallography," VEB Verlag Technik, Berlin, 1970.

29. M. J. Dapino, "Nonlinear and Hysteretic MagnetomechaNical Model for Magnetostrictive Transducers," PhD Thesis, Iowa State University, 1999.

30. E. Clark, J. B. Restorff, M. Wun-Fogle, T. A. Lograsso and D. L. Schlagel, "Magnetostrictive Properties of Bodycentered Cubic Fe-Ga and Fe-Ga-Al Alloys," IEEE Transactions on Magnetics, Vol. 36, No. 5, 2000, pp. 3238- 3240.

31. R. A. Kellog, "Development and Modeling of Iron-Gallium Alloys," PhD Thesis, Iowa State University, 2003.

32. M. J. Dapino, "On Magnetostrictive Materials and their Use in Adaptive Structures," International Journal of Structural Engineering and Mechanics, Vol. 17, No. 3-4, 2004, pp. 303-329.

33. M. J. Dapino, F. T. Calkins and A. B. Flatau, "Magnetostrictive Devices," Wiley Encyclopedia of Electrical and Electronics Engineering, In. J. G. Webster, Ed., John Wiley and Sons Inc., New York, Vol. 12, 1999, pp. 278- 305.

34. J. B. Restorff, "Magnetostrictive Materials and Devices," Encyclopedia of Applied Physics, Vol. 9, 1994, pp. 229- 244.

35. R. Chung, R. Weber and D. Jiles, "A Terfenol-D Based Magnetostrictive Diode Laser Magnetometer," IEEE Transactions on Magnetics, Vol. 27, No. 6, 1991, pp. 5358-5243.

36. M. J. Dapino, R. C. Smith and A. B. Flatau, "StructuralMagnetic Strain Model for Magnetostrictive Transducers," IEEE Transactions on Magnetics, Vol. 36, No. 3, 2000, pp. 545-556.

37. P. R. Downey and A. B. Flatau, "Magnetoelastic Bending of Galfenol for Sensor Applications," Journal of Applied Physics, Vol. 97, No. 10, 2005, pp. (10R505)1-3.

38. J. Atulasimha, A. B. Flatau and R. A. Kellogg, "Sensing Behavior of Varied Stoichiometry Single Crystal Fe-Ga," Journal of Intelligent Material Systems and Structures, Vol. 17, No. 2, 2006, pp. 97-105.

39. G. W. Farnell and E. L. Adler, "Elastic Wave Propagation in Thin Layers," Physical Acoustics: Principles and Methods, In. W. P. Mason and R. N. Thurston Eds., Academic Press, New York, Vol. 9, 1972, pp. 35-127.

40. C. Lardat, C. Maerfeld and P. Tournois, "Theory and Performance of Acoustical Dispersive Surface Wave Delay Lines," Proceedings of the IEEE, Vol. 59, No. 3, 1971, pp. 355-368.

41. V. E. Lyamov, "Polarization Effects and Interaction Anisotropy of Acoustic Waves in Crystals," MSU Publishing, Moscow, 1983.

42. E. Clark, K. B. Hathaway, M. Wun-Fogle, J. B. Restorff, T. A. Lograsso, V. M. Keppens, G. Petculescu and R. A. Taylor, "Extraordinary Magnetoelasticity and Lattice Softening in Bcc Fe-Ga Alloys," Journal of Applied Physics, Vol. 93, No. 10, 2003, pp. 8621-8623.

43. E. Clark, "Structural Fe-based Alloys with High Magnetostriction," Galfenol Workshop, University of Maryland, 29 January 2004.

44. A. Zakharenko, "On Cubic Crystal Anisotropy for Waves with Rayleigh-Wave Polarization," Non-Destructive Testing and Evaluation, Vol. 21, No. 2, 2006, pp. 61-77.

45. T. J. C. Liu and C. H. Chue, "On the Singularities in a Bimaterial Magneto-Electro-Elastic Composite Wedge under Anti-Plane Deformation," Composite Structures, Vol. 72, No. 2, 2006, pp. 254-265.

46. F. Ramirez, P. R. Heyliger and E. Pan, "Free Vibration Response of Two-Dimensional Magneto-Electro-Elastic Laminated Plates," Journal of Sound and Vibration, Vol. 292, No. 3-5, 2006, pp. 626-644.

47. E. Clark, "Magnetostrictive Rare Earth-Fe2 Compounds," Ferromagnetic Materials, In. E. P. Wohlfarth, Ed., North-Holland Publishing Company, Amsterdam, Vol. 1, Chapter 7, 1980, pp. 531-589.

48. V. I. Al'shits and V. N. Lyubimov, "On Existence of Piezoelectric and Piezomagnetic Properties of Anisotropic Media," Soviet Kristallographia, Moscow, Vol. 35, No. 2, 1990, pp. 483-484.

49. D. Royer and E. Dieulesaint, "Elastic Waves in Solids I," Springer-Verlag, New York, 2000.

50. Y. V. Gulyaev, I. E. Dikshtein and V. G. Shavrov, "Magnetoacoustic Surface Waves in Magnetic Crystals at Reorientation Phase Transitions," Uspekhi Phizicheskikh Nauk, Moscow, Vol. 167, No. 7, 1997, pp. 735-750.

CHAPTER 9

Topological Design of Cellular Phononic Band Gap Crystals

Yang Fan Li [1], Xiaodong Huang [1,2,*] and Shiwei Zhou [1]

[1] Centre for Innovative Structures and Materials, School of Engineering, RMIT University, GPO Box 2476, Melbourne 3001, Australia

[2] Key Laboratory of Advanced Technology for Vehicle Body Design & Manufacture, Hunan University, Changsha 410082, China

ABSTRACT

This paper systematically investigated the topological design of cellular phononic crystals with a maximized gap size between two adjacent bands. Considering that the obtained structures may sustain a certain amount of static loadings, it is desirable to ensure the optimized designs to have a relatively high stiffness. To tackle this issue, we conducted a multiple objective optimization to maximize band gap size and bulk or shear modulus simultaneously with a prescribed volume fraction of solid material so that the resulting structures can be lightweight, as well. In particular, we first conducted the finite element analysis of the phononic band gap crystals and then adapted a very efficient optimization procedure to resolve this problem based on bi-directional evolutionary structure optimization (BESO) algorithm in conjunction with the homogenization method. A number of optimization results for maximizing band gaps with bulk and shear modulus constraints are presented for out-of-plane and in-plane modes. Numerical results showed that the optimized structures are similar to those obtained for composite case, except that additional slim connections are added in the cellular case to support the propagation of shear wave modes and meanwhile to satisfy the prescribed bulk or shear modulus constraints.

KEYWORDS

cellular phononic crystals; band gap; homogenization; BESO; bulk modulus; shear modulus

1. INTRODUCTION

The propagation of mechanical waves in periodic structures has aroused a growing research interest in recent years. The existence of band gaps in phononic crystals can lead to a variety of potential applications in sound insulation, shock isolations, seismic wave-proofing, acoustic wave filtering, waveguides, negative refraction, *etc.* [1]. In phononic crystals (PnCs), the elastic properties are periodic functions of position with a periodicity comparable to the wavelength of the corresponding wave field [2]. The underlying mechanism for the formation of band gaps relies on the different scattering of a mechanical wave at the interface between constituent phases with high contrast in density and/or elastic constants. In the past few years, most studies consider PnCs consisting of two solid components, solid-fluid or fluid-fluid components. Among these different patterns, cellular PnCs with air or vacuum cylindrical holes embedded in a host material are more economical and practical, as well as multi-functional. Compared with composite PnCs, cellular PnCs can be very light and achieve high toughness, while reducing the fabrication cost. In this research we only consider the case of cellular phononic band gap crystals.

Calculation of band structure for acoustic or elastic waves propagating in periodic composite structures was first carried out by Sigalas *et al.* in 1992 [3] by using plane wave expansion method (PWE), and the concept of PnCs was first conceived by Kushwaha in 1993 [4,5]. Although the PWE method is capable to calculate dispersion relation of mechanical waves, it has convergence and accuracy problems in the case of fluid/solid system, especially for the case when the contrast of material parameters between inclusions and the substrate is very high [2]. Later, various techniques have been successfully introduced to calculate band structure of acoustic and elastic waves in fluid/solid or vacuum/solid PnCs, such as the transfer matrix method (TMM) [6], multiple scattering theory (MST) [7], finite-difference time-domain (FDTD) [8,9,10], Rayleigh method (RM) [11], and finite element method (FEM) [12], *etc.*

The application of phononic band gap crystals highly depends on the width of the band gap and the key issue is therefore to engineer the phononic band gap as wide as possible. A topology optimization method provides an effective means for systematic and scientific approach of designing phononic structures with an optimal band gap. Optimization of phononic band gap crystals was first conducted by Sigmund and Jensen [13] based on finite element method (FEM) in combination with the method of moving asymptotes (MMA). In the following, genetic algorithm (GA) and gradient-based topology optimization, in conjunction with FEM or the fast plane wave expansion method (FPWE), are developed to maximize the band gap sizes of phononic band gap crystals [14,15,16,17,18,19,20]. However, most of the optimization work is conducted for the case of solid composite materials, and few studies focus on the case of air/vacuum inclusions embedded in a host medium [16]. Dong *et al.* [17] conducted a multi-objective optimization of 2D porous phononic crystals for maximizing band gap width and minimizing mass of structure simultaneously by using non-dominated sorting-based genetic algorithm II (NSGA-II). To guarantee the self-support of cellular phononic crystals, an artificial geometrical constraint is adapted.

It is apparent that current research in this area is insufficient and further systematic investigation into the design of cellular phononic band gap crystals is necessary, especially for their multi-functionalities. In order to take advantage of light-weight cellular phononic crystals, mass of structure should be minimized or subjected to a certain constraint. In fact, both optimization results for composite phononic crystals and the existing cellular results indicate that maximizing band gap tends to be the isolation of solid material with high density and stiffness; however, transverse/shear wave is not supported in air/vacuum. Thus, it is meaningful to make sure that the optimized phononic band gap structures have continuous distribution of solid material to sustain static loads. In contrast with simply adding a geometrical constraint to the optimized structures, it is more meaningful to consider the additional functionality of phononic crystals, such as bulk or shear modulus, so that the resulting structures could achieve appropriate stiffness for undergoing external forces. To the authors' best knowledge, no work has been reported yet to optimize the cellular phononic crystals with a stiffness constraint and volume constraint simultaneously.

The purpose of this research is to maximize the gap size between two appointed bands for cellular phononic crystals subject to bulk or shear modulus constraint with a given volume fraction. The rest of the paper is organized as follows: InSection 2 we introduce the essential governing equations and related theories of phononic crystals, along with the homogenization theory to calculate the effective bulk or shear modulus of cellular structures. Then we formulate the optimization problem into the mathematical equations and present the topology optimization algorithm used in this paper. In Section 3, we discuss the numerical implementations of the optimization problem and the associated sensitivity analysis. A number of numerical examples and optimized cellular phononic structures are presented in Section 4 for out-of-plane waves, in-plane waves, and the coupled in-plane and out-of-plane waves, respectively. This is followed by the conclusions.

2. THEORY AND OPTIMIZATION PROBLEM

2.1. Band Gap Analysis of Phononic Crystals

The governing equation of mechanic waves that propagate in a heterogeneous medium is given by:

$$\rho(\mathbf{r})\ddot{\mathbf{u}} = \nabla[\lambda(\mathbf{r}) + 2\mu(\mathbf{r})](\nabla\cdot\mathbf{u}) - \nabla\times[\mu(\mathbf{r})\nabla\times\mathbf{u}] \quad (1)$$

where λ and μ denote the Lame's coefficients; ρ is the material density; u={ux,uy,uz}T is the displacement vector, and r = (*x*, *y*) denotes the position vector. In this paper, we consider two dimensional phononic crystals with square lattice and assume the propagation of elastic waves is restricted to the *x*–*y* plane only. Noting that the wave field is independent ofz *i.e.*, ∂u/∂z=0, Equation (1) can be split into two equations which govern the in-plane longitudinal and transverse waves (which is called in-plane wave mode), and one out-of-plane equation that governs the out-of-plane waves (which is called out-of-plane wave mode) as:

$$\rho(\mathbf{r})\frac{\partial^2 u_x}{\partial t^2} = \frac{\partial}{\partial x}\left[(\lambda(\mathbf{r}) + 2\mu(\mathbf{r}))\frac{\partial u_x}{\partial x} + \lambda(\mathbf{r})\frac{\partial u_y}{\partial y}\right] + \frac{\partial}{\partial y}\left[\mu(\mathbf{r})\left(\frac{\partial u_x}{\partial y} + \frac{\partial u_y}{\partial x}\right)\right] \tag{2}$$

$$\rho(\mathbf{r})\frac{\partial^2 u_y}{\partial t^2} = \frac{\partial}{\partial x}\left[\mu(\mathbf{r})\left(\frac{\partial u_x}{\partial y} + \frac{\partial u_y}{\partial x}\right)\right] + \frac{\partial}{\partial y}\left[\lambda(\mathbf{r})\frac{\partial u_x}{\partial x} + (\lambda(\mathbf{r}) + 2\mu(\mathbf{r}))\frac{\partial u_y}{\partial y}\right] \tag{3}$$

$$\rho(\mathbf{r})\frac{\partial^2 u_z}{\partial t^2} = \frac{\partial}{\partial x}\left[\mu(\mathbf{r})\frac{\partial u_z}{\partial x}\right] + \frac{\partial}{\partial y}\left[\mu(\mathbf{r})\frac{\partial u_z}{\partial y}\right] \tag{4}$$

According to Bloch's theorem, for periodic structures the displacement vector $\mathbf{u}(\mathbf{r})$ should satisfy the form:

$$\mathbf{u}(\mathbf{r}, \mathbf{k}) = \mathbf{u}_k(\mathbf{r})\, e^{i(\mathbf{kr})} \tag{5}$$

where $\mathbf{u}_k(\mathbf{r})$ is a periodic function of $\mathbf{r}$ with the same periodicity to the structure. $\mathbf{K} = (k_x, k_y)$ is the Bloch wave vector [21] By inserting Equation (5) into either Equations (2) and (3) or Equation (4), the governing equations can be converted to two eigenvalue problems for in-plane and out-of-plane waves, respectively. We could easily solve the problem by the finite-element method and both eigenvalue problems can be written as:

$$\left(\mathbf{K}(\mathbf{k}) - \omega(\mathbf{k})^2\mathbf{M}\right)\mathbf{u} = 0 \tag{6}$$

where eigenvectors $\mathbf{u} = \mathbf{u}_k(\mathbf{r})$. $\mathbf{K}$ and $\mathbf{M}$ are the stiffness matrix and mass matrix, respectively. It should be noted that, due to the symmetry of the unit cell, the above eigenvalue equations only need to be solved within the first Brillouin zone. Moreover, it has been verified that the searching area can be further reduced to the wave vectors on the boundary of the irreducible Brillouin zone for the calculation of band structures [22]. For a 2D phononic crystal with the square lattice shown in Figure 1a, the boundary of the irreducible first Brillouin zone is sides of the triangle *Γ-X-M-Γ* shown in Figure 1b. In the following numerical examples, the wave vectors **k** = (k_x, k_y) are appointed with 11 equally-spaced points along each boundary and start from the point *Γ* (0, 0), to *X*(π/*a*, 0), then *M*(π/*a*, π/*a*), and finally come back to the point *Γ*.

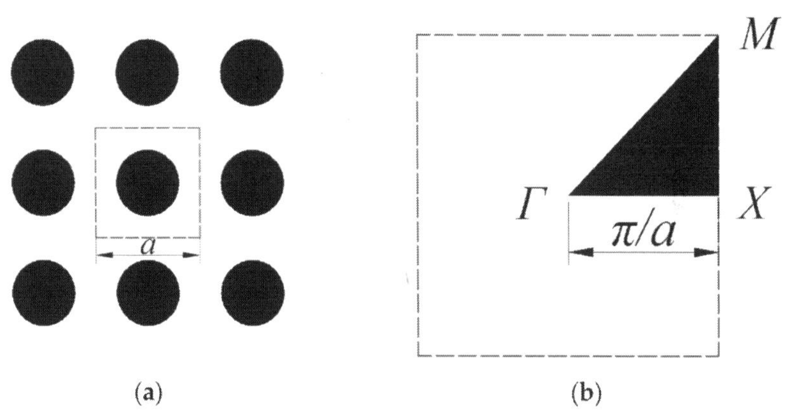

Figure 1. (**a**) Phononic crystals with 3 × 3 unit cells; and (**b**) irreducible first Brillouin zone (*Γ-X-M-Γ*).

2.2. Static Effective Stiffness of Phononic Crystals

When cellular phononic crystals are considered as a component material for a device, the periodic unit cell would be very small compared with the size of the structural body. Therefore, the static effective elasticity tensor of cellular phononic crystals can be calculated by homogenization theory [23,24] in terms of the material distribution in the unit cell, which is stated as:

$$E_{ij}^{H} = \frac{1}{|Y|}\int_{\Omega}\left(\left\{\varepsilon_0^i\right\} - \left\{\varepsilon^i\right\}\right)^T [E]\left(\left\{\varepsilon_0^j\right\} - \left\{\varepsilon^j\right\}\right) d\Omega \quad (7)$$

where EHij is homogenized elasticity tensor, *[E]* is the constitutive matrix at a given point, |Y| denotes the area of the unit cell Ω, i, j = 1, 2, 3 for two dimensional inhomogeneous structures, {εi0} are three linear independent test strain fields as {ε10}={1,0,0}, {ε20}={0,1,0}, {ε30}={0,0,1}, {εi} are the introduced strain fields, which are the solutions to the standard finite element equation and subject to periodic boundary condition and the test strain fields {εi0}.

The static stiffness of cellular structures is usually described by bulk modulus κ^H or shear modulus G^H and can be expressed in terms of the components of effective elasticity tensor EHij as:

$$\kappa^H = \frac{1}{4}\left(E_{11}^H + E_{12}^H + E_{21}^H + E_{22}^H\right) \tag{8}$$

$$G^H = E_{33}^H \tag{9}$$

To design a stiffer cellular phononic band gap crystal, the best way is to maximize bulk or shear modulus and the band gaps simultaneously. However, our numerical experiences show that it is impossible to achieve two goals at the same time, since the optimal solutions to maximize bulk or shear modulus and maximize phononic band gaps are in two opposite directions. An alternate way is to make a compromise by adding a bulk or shear modulus constraint to the optimization problem of maximizing band gaps, so that the resulting structures can possess a relatively big band gap while also maintaining a higher stiffness. In the following examples, dimensionless stiffness constraints are used instead of effective bulk or shear modulus. Specifically, $\kappa = \kappa^H/\kappa_0$ and $G = G^H/G_0$ are used as static effective bulk and shear modulus constraints, where κ_0 and G_0 are the bulk and shear moduli, respectively, of the solid material.

For a cellular structure with a volume fraction V_f of solid material in the design domain, the maximum value of bulk or shear modulus should satisfy the Hashin–Shtrikman bounds for two-phase materials [25]. The corresponding dimensionless upper bounds of stiffness are given as below:

$$\kappa_{upper} = \frac{V_f G_0}{\left(1 - V_f\right) \kappa_0 + G_0} \tag{10}$$

$$G_{upper} = \frac{V_f \kappa_0}{\left(1 - V_f\right) \left(\kappa_0 + 2G_0\right) + \kappa_0} \tag{11}$$

In the optimization of cellular phononic crystals, the stiffness constraint value should be not greater than κ_{upper} andG_{upper}, otherwise the optimization will tend to maximize the bulk or shear modulus other than band gaps. Thus, the ratio ofκ/κ_{upper} and G/G_{upper} should be carefully chosen as the stiffness constraint. We will discuss this later in the numerical examples.

2.3. Optimization Problem

The goal of this paper is to design cellular phononic crystals with a desirable band gap by properly rearranging the distribution of air holes and solid material subject to a given material volume and certain stiffness. Considering the absence of the essential length scale in the governing equations, here we choose to optimize the relative band gap size, which is characterized by the gap-midgap ratio. To optimize the relative band gap size between the nth and $(n + 1)^{th}$ band for in-plane waves or out-of-plane waves, the optimization problem can be mathematically formulated with objective and constraint functions as follows:

$$\text{Maximize}: \quad f\left(x_e\right) = 2\frac{\min \omega_{n+1}\left(\mathbf{k}\right) - \max \omega_n\left(\mathbf{k}\right)}{\min \omega_{n+1}\left(\mathbf{k}\right) + \max \omega_n\left(\mathbf{k}\right)} \tag{12}$$

$$\text{Subject to :} \qquad \kappa \geqslant \kappa^* = \beta\kappa_{upper} \quad or \quad G \geqslant G^* = \beta G_{upper} \tag{13}$$

$$V_f^* = \sum_{e=1}^{N} x_e V_e \quad x_e = x_{\min} \text{or} 1 \tag{14}$$

where the objective function $f(x_e)$ denotes the gap-midgap ratio between the nth and $(n + 1)^{th}$ bands which is defined by the percentage in the following band diagrams; ω_n, ω_{n+1} are eigenfrequecies at target bands; κ^* and G^* are static effective bulk and shear modulus constraint, and β is the ratio of stiffness constraint over its upper bound value at a same volume. Obviously the magnitude of β is located in the range between 0 and 1; $V*f$ is the volume constraint; x_e is the artificial design variable, which denotes the material type (air or solid material) for each element.

2.4. Bi-Directional Evolutionary Structural Optimization (BESO)

Here, we employ the bi-directional evolutionary structural optimization (BESO) method, a very efficient and effective topology algorithm in optimum material distribution problems for continuum structures, to resolve the optimization problem defined by Equations (12)–(14). BESO is a further developed version of evolutionary structural optimization (ESO), which was originally proposed by Xie and Steven [26,27] in the early 1990s. The basic concept of ESO is to gradually remove low efficient materials from the structure so that the rest part evolves to an optimum. BESO allows adding materials to the most efficient regions as well as deleting insufficient ones [28,29]. In the present paper, we use the new BESO method which is proposed by Huang and Xie and greatly improved the robustness and computational efficiency of the original ESO and BESO methods [30,31,32]. It has also been demonstrated that the current BESO method is well capable of the design of periodic microstructures for cellular materials and composites, such as the photonic crystals [33,34,35,36].

The problem of maximizing the band gap size for phononic crystals is in essence how to arrange the spatial distributions of the solid material and air within the unit cell. In other words, the topology optimization problem can be transferred to alter the artificial design variable of each element after the finite element discretization. In order to get the gradient information of object function with respect to the design variables, it is necessary to interpolate the material properties between air and the solid material. In the optimization of composite phononic crystals, which consisted of two solid materials, a simple linear material interpolation scheme works very well [13]. However, artificial localized modes are often encountered in low density regions for the case of optimization of cellular phononic crystals due to the extremely high contrast of mass and stiffness between air and the solid material. To avoid this problem, we apply a similar material interpolation scheme with penalization as in the studies on the topology optimization of continuum structures for natural frequencies [37]. The interpolation scheme is given as:

$$\rho(x_e) = x_e \rho_0 \tag{15}$$

$$E(x_e) = \left[\frac{x_{\min} - x_{\min}^p}{1 - x_{\min}^p} \left(1 - x_e^p\right) + x_e^p \right] E_0 \quad (0 < x_{\min} \leqslant x_e \leqslant 1) \tag{16}$$

where ρ_0 and E_0 represent the density and Young's modulus of solid material, respectively; p is the penalty exponent; $x_e$stands for a design variable, $x_e = x_{min}$ denotes element e is composed of air, and $x_e = 1$ means element e is composed of solid material. To avoid singularity in finite element analysis, x_{min} in the calculation is usually set to be a very small value that is slightly larger than 0. In the following example, the value is chosen as $x_{min} = 1 \times 10^{-6}$. Current material interpolation scheme for different p with x_{min} =

0.01 is plotted in Figure 2. It is clearly seen that when x_e approaches the ends 0 or 1, the current model is same as the linear material interpolation case.

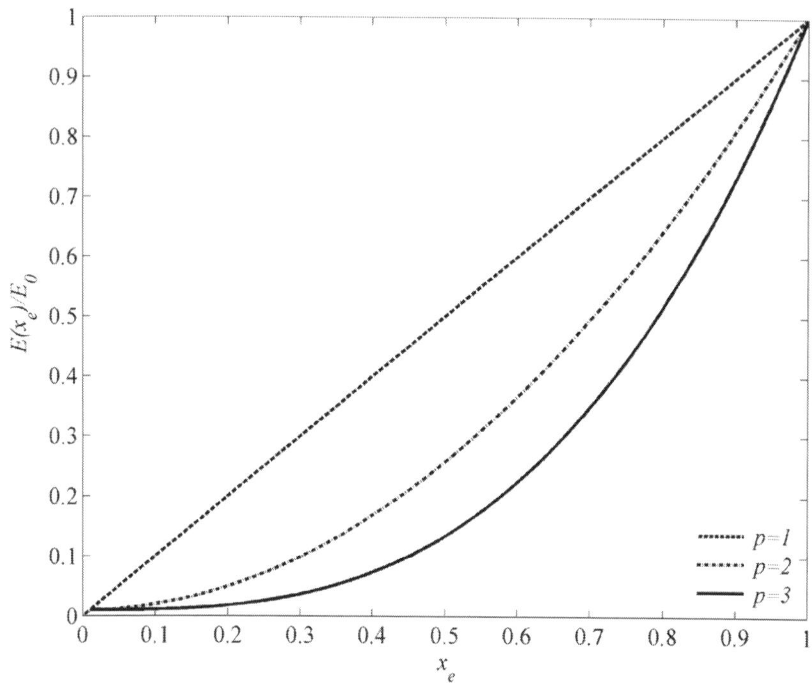

Figure 2. Material interpolation scheme with $x_{min} = 0.01$.

It should be pointed out that the discrete design variable x_e is set to be $x_{\min}$ or 1 only in the traditional BESO method [32,37]. However, the optimization of phononic crystals is very sensitive to the variation of design variable, so the change of design variable in each iteration is limited to be $\Delta x_e = 0.1$ to stabilize the optimization process. In other words, x_e can take 11 discrete values, as xe∈{xmin,0.1,···,0.9,1}. Our numerical experience in the optimization of composite case shows that the discrete scheme could also lead to a clear solid/void final topology.

3. NUMERICAL IMPLEMENTATION AND BESO PROCEDURE

3.1. Reformulation of Objective Function

Current optimization problem stated in Equations (12)–(14) has multiple constraints, including a stiffness constraint and a volume constraint. Like the optimization problem of continuum structures with an additional displacement constraint using ESO method [38], the stiffness constraint can be added by introducing a Lagrangian multiplier. The objective function can be modified for the bulk modulus constraint to:

$$\text{Maximize}: \qquad f(x_e) = 2\frac{\min\omega_{n+1}(\mathbf{k}) - \max\omega_n(\mathbf{k})}{\min\omega_{n+1}(\mathbf{k}) + \max\omega_n(\mathbf{k})} + \lambda(\kappa - \kappa^*) \qquad (17)$$

and for the shear modulus constraint to:

$$\text{Or}: \qquad f(x_e) = 2\frac{\min\omega_{n+1}(\mathbf{k}) - \max\omega_n(\mathbf{k})}{\min\omega_{n+1}(\mathbf{k}) + \max\omega_n(\mathbf{k})} + \lambda(G - G^*) \qquad (18)$$

When the bulk or shear modulus is equal to the prescribed ones, the above equations are equivalent to the original objective function defined in Equation (13). Otherwise, taking the bulk modulus constraint case for example, if $\kappa > \kappa^*$, which means the constraint is already satisfied and λ is set to be 0; if $\kappa < \kappa^*$, which means the constraint is not satisfied yet and we need to maximize the bulk modulus first, so λ needs to be infinity. The case of shear modulus constraint is the same as that of bulk modulus. The determination of Lagrangian multiplier will be discussed later. After this modification, optimization of cellular phononic crystals with stiffness constraint can be solved as the standard band gap optimization problem.

3.2. Sensitivity Analysis

The gradient of objective function with respect to the change of design variable x_e defined in Equation (17) for bulk modulus constraint case can be calculated by:

$$\alpha = \frac{\partial f(x_e)}{\partial x_e} = 4\frac{\max\omega_n(\mathbf{k})\frac{\partial\min\omega_{n+1}(\mathbf{k})}{\partial x_e} - \min\omega_{n+1}(\mathbf{k})\frac{\partial\max\omega_n(\mathbf{k})}{\partial x_e}}{(\min\omega_{n+1}(\mathbf{k}) + \max\omega_n(\mathbf{k}))^2} + \lambda\frac{\partial\kappa}{\partial x_e} \quad (19)$$

and for shear modulus constraint case by:

$$\alpha = \frac{\partial f(x_e)}{\partial x_e} = 4\frac{\max\omega_n(\mathbf{k})\frac{\partial\min\omega_{n+1}(\mathbf{k})}{\partial x_e} - \min\omega_{n+1}(\mathbf{k})\frac{\partial\max\omega_n(\mathbf{k})}{\partial x_e}}{(\min\omega_{n+1}(\mathbf{k}) + \max\omega_n(\mathbf{k}))^2} + \lambda\frac{\partial G}{\partial x_e} \quad (20)$$

where the sensitivities of eigenfrequencies can be expressed by the following equation with the assumption that eigenvectors are nomalized to the global mass matrix for a given wave vector **k**:

$$\frac{\partial\omega_n}{\partial x_e} = \frac{1}{2\omega_n}\mathbf{u}(\mathbf{k})_n^T\left(\frac{\partial\mathbf{K}}{\partial x_e} - \omega_n^2\frac{\partial\mathbf{M}}{\partial x_e}\right)\mathbf{u}(\mathbf{k})_n \quad (21)$$

where $\mathbf{u}(\mathbf{k})_n$ is the corresponding eigenvector; **K** and **M** are the elemental stiffness and mass matrix, respectively.

According to the relations in Equations (8) and (9), the sensitivities of homogenized bulk or shear modulus can be computed as:

$$\frac{\partial\kappa^H}{\partial x_e} = \frac{1}{4}\left(\frac{\partial E_{11}^H}{\partial x_e} + \frac{\partial E_{12}^H}{\partial x_e} + \frac{\partial E_{21}^H}{\partial x_e} + \frac{\partial E_{22}^H}{\partial x_e}\right) \quad (22)$$

$$\frac{\partial G^H}{\partial x_e} = \frac{\partial E_{33}^H}{\partial x_e} \quad (23)$$

Using the adjoint varibale method [39], the sensitivity of the homogenized elasticity tensor with respect to design variable can be derived as:

$$\frac{\partial E_{ij}^H}{\partial x_e} = \frac{1}{|Y|}\int_\Omega \left(\left\{\varepsilon_0^i\right\} - \left\{\varepsilon^i\right\}\right)^T \frac{\partial E_{ij}}{\partial x_e}\left(\left\{\varepsilon_0^j\right\} - \left\{\varepsilon^j\right\}\right) d\Omega \tag{24}$$

Considering the material interpolation scheme defined in Equation (16), the sensitivity of the homogenized elasticity tensor can be written as:

$$\frac{\partial E_{ij}^H}{\partial x_e} = \frac{p x_e^{p-1}}{|Y|}\frac{1 - x_{\min}}{1 - x_{\min}^p}\int_\Omega \left(\left\{\varepsilon_0^i\right\} - \left\{\varepsilon^i\right\}\right)^T E_{ij}^0\left(\left\{\varepsilon_0^j\right\} - \left\{\varepsilon^j\right\}\right) d\Omega \tag{25}$$

where E0ij is the elasticity tensor of solid material.

Up to now, the only unknown parameter in Equation (19) or Equation (20) is the Lagrangian multiplier, which highly influences the relative ranking of the overall sensitivity of the objective function.

3.3. Determination of the Lagrangian Multiplier

Before the calculation of the overall sensitivity of the objective function, the Lagrangian multiplier λ should be determined first. Here we introduce an intermediate parameter *w* in our program, which is defined by:

$$\lambda = \frac{1 - w}{w} \tag{26}$$

where *w* is a constant value which can vary from a extremely small value $w_{\min}$ e.g., 10^{-20}, to 1, so the resulting Lagrangian multiplier λ can be located in the range of 0 to infinity.

Initially, the search for the proper value of w starts from $w = 1$ with a lower bound $w_{lower} = w_{min} = 10^{-20}$ and an upper bound $w_{upper} = w_{max} = 1$. Then, with the Lagrangian multiplier computed by Equation (26), the sensitivity of objective function is calculated by Equation (19) or Equation (20). Then, based on the ranking of the sensitivity the design variables are updated to satisfy the volume fraction in the next iteration. The bulk or shear modulus in the next iteration can be approximately estimated by the variation of bulk or shear modulus and the value in the current iteration, κ^i or G^i, as:

$$\kappa^{i+1} \approx \kappa^i + \sum_e \frac{d\kappa}{dx_e} \Delta x_e \tag{27}$$

or:

$$G^{i+1} \approx G^i + \sum_e \frac{dG}{dx_e} \Delta x_e \tag{28}$$

where $\frac{d\kappa}{dx_e}$ and $\frac{dG}{dx_e}$ are computed by Equations (22) and (23). After that, the estimated bulk or shear modulus will be compared with the current bulk or shear modulus constraint. If $\kappa^{i+1} < \kappa^*$, which means the current constraint is not satisfied and current Lagrangian multiplier λ is too small, we would update w with a smaller value by:

$$\hat{w} = \frac{w + w_{lower}}{2} \tag{29}$$

Meanwhile, the upper bound of w_{upper} will be moved to w. If $\kappa^{i+1} > \kappa^*$, which means the current constraint is already satisfied, still we would update w, however, with a larger value by:

$$\hat{w} = \frac{w + w_{upper}}{2} \tag{30}$$

and the lower bound w_{lower} will be moved to w. The Lagrangian multiplier is updated at the same time. This procedure will be repeated until the difference between w_{upper} and w_{lower} is less than 10^{-15}. Finally the appropriate Lagrangian multiplier is obtained after several iterations.

3.4. BESO Procedure

In the optimization of composite phononic band gap crystals, our numerical experiences show that the filter scheme will lead to more reliable and steady evolution history. Therefore, we continue to apply a filter by averaging the elemental sensitivity with its neighbor elements in the optimization of cellular phononic crystals. Instead of averaging the overall sensitivity calculated by the Equation (19) or Equation (20), the filter is separately applied to the sensitivity of gap-midgap ratio and homogenized bulk or shear modulus since our numerical simulation indicates that filtering the sensitivity separately could better avoid the checkerboard problem [40]. The filter scheme refers to [37].

The detailed optimization procedure using BESO method can be illustrated by the Figure 3. We choose a simple or random unit cell as the initial design and discretize the design domain by a finite element mesh. Then we define the objective volume fraction of the solid material V∗f, evolutionary ratio *ER* and penalty p in the material interpolation scheme. In the following step, we first conduct finite element analysis for several discrete wave vectors **k** along the first Brillouin zone with Bloch boundary conditions to get corresponding eigenvalues and eigenvectors to Equation (6). After this, the sensitivities of gap-midgap ratio for each element are calculated and filtered. Then we use homogenization method to get the effective bulk or shear modulus (κ^i or G^i) of current configuration. Additionally, the sensitivities of homogenized parameters are calculated and filtered. Before the update of the

Lagrangian multiplier of the constraint, the target volume in the next iteration is defined as:

$$V_f^{i+1} = V_f^i(1 - ER) \quad when \quad V_f^i > V_f^* \tag{31}$$

where i denotes the iteration number. Once the final volume fraction V∗f is achieved, the volume will be kept as a constant. At the same time, corresponding bulk modulus constraint κ^* or shear modulus constraint G^* at the target volume fraction in next iteration is computed by Equation (10) or Equations (11) and (13) with a given constraint ratio β. ER = 2%, V∗f = 50% and β = 0.3 are used throughout this paper, unless otherwise stated. Thereafter, the appropriate Lagrangian multiplier in the next iteration is determined by the procedure illustrated in previous section. Then the overall sensitivity of the objective function can finally be obtained.

Based on the relative ranking of the calculated overall sensitivity, a threshold of the sensitivity number, αith, is determined by using the bi-section method so that the volume fraction in the next iteration is equal to Vi+1f. The design variable of each element is modified by comparing its sensitivity number αie with the threshold as:

$$x_e^{i+1} = \begin{cases} \min(x_e^i + \Delta x, 1), & \text{if } \alpha_e^i > \alpha_{th}^i \\ \max(x_e^i - \Delta x, 0), & \text{if } \alpha_e^i < \alpha_{th}^i \end{cases} \tag{32}$$

where Δx=0.1 throughout the paper, which means BESO uses discrete design variables.

The above procedure is repeated until the convergence criteria are satisfied. It should be pointed out that the stiffness constraint in the optimization process is not a constant value like volume constraint, and would evolve with the volume fraction for the sake of stability of the evolutionary history. Additionally, although discrete intermediate design variables are used during the optimization process, the optimized design we obtained can naturally converge to an almost solid/void design due to the adoption of the material interpolation scheme.

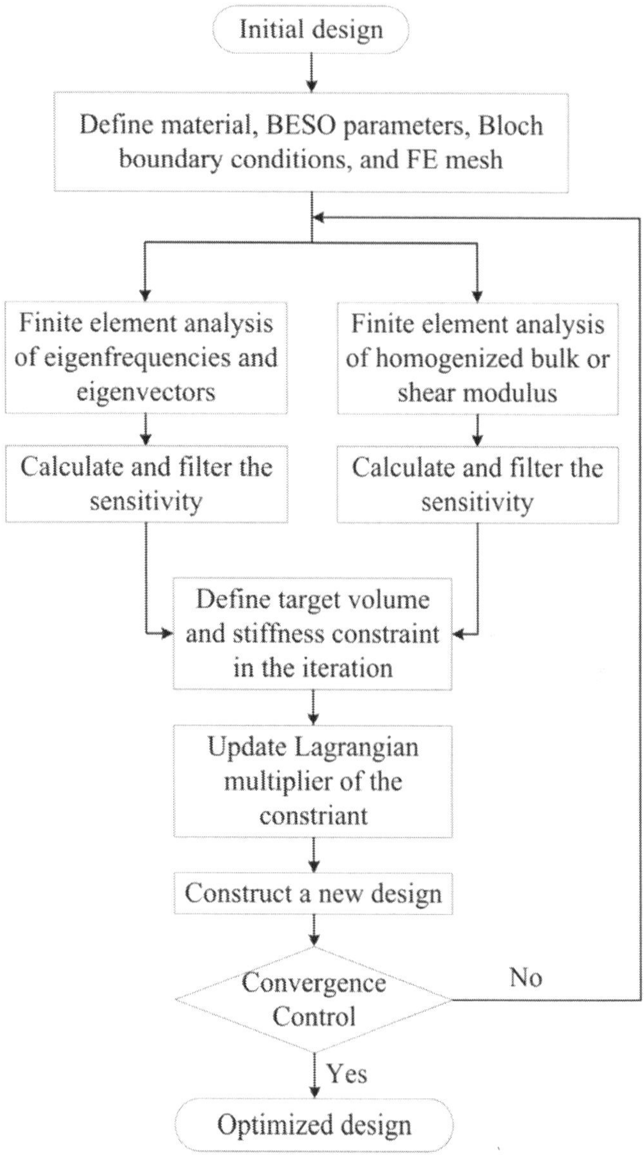

Figure 3. Flow chart of optimization procedure using BESO method.

4. RESULTS AND DISCUSSION

To demonstrate the capability of the proposed algorithm, this section will present a number of band gap optimization results we obtained for the out-of-plane mode, in-plane mode, and the combined mode of these two with bulk and shear modulus constraints, respectively. In the following examples, we consider the design of a 2D cellular phononic crystal with square lattice. The objective is to maximize the relative band gap size between two adjacent bands. All the numerical simulations are conducted with a volume constraint $V*f = 50\%$.

Silicon is used as the solid material due to its popularity in the design of photonic crystals, which are the electromagnetic analog of phononic crystals and have been extensively studied. The physical properties of silicon are given as ρ = 2330 kg/m^3, λ = 85.502 GPa, and μ = 72.835 GPa [17]. The unit cell with dimensionless lattice length $a = 1$ is discretized into 64 × 64 linear four-node elements. All of the evolutionary rates of volume fraction used in the BESO procedure are ER = 2%. The filter radius for the sensitivity of band gap is $r_{min1} = 2\sqrt{a}/30$ while for the sensitivity of homogenized bulk or shear modulus is $r_{min2} = 2\sqrt{a}/40$. The eigenfrequencies (ω) in the band structures are normalized by $\omega a/2\pi C$, where $C = 340$ m/s denotes the wave speed in air.

4.1. Out-of-Plane Mode

4.1.1. Influence of Stiffness Constraint

As mentioned in Section 2.2, the value of stiffness constraint directly affects the final optimization results of cellular phononic band gap crystals. To investigate the influence of different bulk and shear modulus constraint on the optimized band gap size, we conduct a series of optimizations starting from a same initial design with the constraint ratio β defined in Equation (13), changing from 0.1 to 0.7. The relations of optimized relative band gaps and constraint ratio β are plotted inFigure 4 along with the corresponding final topologies consisting of 3 × 3 unit cells. The optimized unit cell is shown in the red dashed box. It can be clearly seen that the optimal relative band gaps size for both bulk and shear modulus constraint case declines continuously with the increase of constraint ratio while the shapes inside the unit cell does not vary a lot. The final relative band gap sizes with shear modulus constraint are slightly bigger than those with bulk modulus constraint but the differences are quite small compared with their net value.

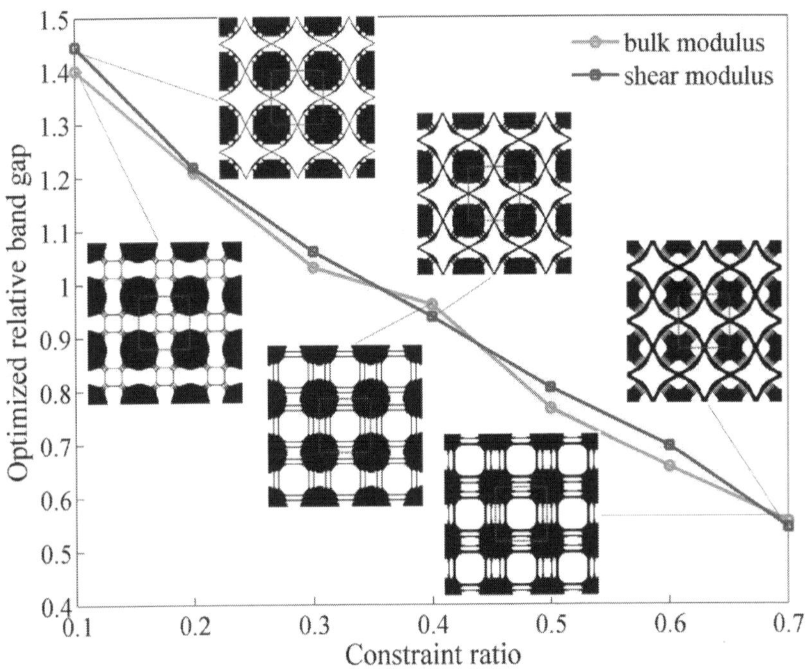

Figure 4. Optimized relative band gap against different constraint ratio with bulk and shear modulus constraints.

The maximum relative band gap size we obtained is up to 144.41% with the shear modulus constraint as $\beta_G = 0.1$, which breaks the record value in the literature [16,17]. This number decreases dramatically by around 90% to 54.27% when the shear modulus constraint ratio β_G increases from 0.1 to 0.7. Meanwhile the corresponding topology with smaller shear modulus constraint tends to have many tiny and slim connections, and these connections gradually disappear or become thicker when the constraint ratio β_G is growing.

The situation for bulk modulus constraint case is very similar. The maximum relative band gap size with the bulk modulus constraint $\beta_\kappa = 0.1$ is 140.02% while the number drops to 55.41% as the bulk modulus constraint increases to $\beta_\kappa = 0.7$. Compared with the topologies we obtain with a shear modulus constraint, optimized structures in this group have different slim connections, while the main shapes of the solid material in the unit cell remain in similar positions.

Figure 4 indicates that the increase of stiffness constraint ratio will lead to a significant drop of optimized relative band gap size. As the bulk or shear modulus constraint gets bigger, the connections in the corresponding resulting topologies become thicker, which are favorable for future manufacture. To make a compromise between the band gap size and stiffness, the stiffness constraint ratio is set as β_κ or $\beta_G = 0.3$ for bulk and shear modulus constraint cases in the following examples.

As mentioned above, we adopt an evolutionary stiffness constraint in the optimization process. As an illustration example, the evolution histories of effective bulk modulus and bulk modulus constraint in each iteration step and the corresponding HS upper bound are plotted in Figure 5 for the optimization case with constraint ratio $\beta_\kappa = 0.3$. At the beginning, the effective bulk modulus is almost equal to the HS upper bound value, which means BESO starts from an initial configuration that is almost consisted of only solid materials. Based on the relation between the effective bulk modulus and the corresponding constraint value, the whole process can be divided into two stages. At the first stage, the bulk modulus constraints are far less than the effective bulk moduli calculated for the current configurations, indicating in these iterations the Lagrangian multiplier is equal to 0 and the objective is only to maximize the relative band gap size. It is observed that the increasing of the band gap in the first stage leads to a dramatic drop of the effective bulk modulus which, once again, reveals that the optimization trends for these two goals are opposite. At the second stage, the Lagrangian multiplier begins to play an important role in the balance of the band gap optimization and the effective bulk modulus optimization. It can be clearly seen that the effective bulk modulus keeps almost the same value as the bulk modulus constraint in the following optimization.

The variation of the volume fraction, band gap size and topology of the unit cell for the same case during the whole optimization process is presented in Figure 6. We start from a simple unit cell, with centered and cornered holes, and the initial volume fraction V_f is approximately equal to 1, which is in consistent with the initial effective bulk modulus. Initially the gap size is a negative value, indicating that there is no band gap between the first and second band. When the volume fraction steadily drops to the predefined constraint value 50%, the band gap size first gradually increases from a negative value to 0, and then rapidly rises after the opening of the band gap. When the Lagrangian multiplier begins to work, the band gap size experiences a fluctuating period and finally achieves its maximum value after the volume fraction is fixed. Meanwhile, the unit cell evolves to an optimal state with almost no grey elements. It takes less than 50 steps for this case to converge to an optimized structure. Furthermore, the optimized configuration is similar to the result reported in the literature [16].

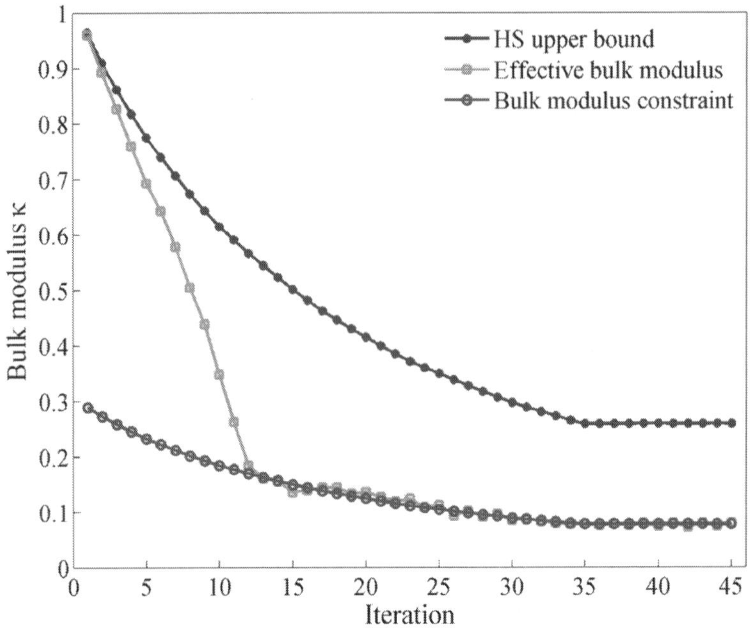

Figure 5. Evolution histories of the effective bulk modulus, bulk modulus constraint and corresponding HS upper bound in the optimization process with constraint ratio$\beta_\kappa = 0.3$.

4.1.2. Out-of-Plane Results with Bulk Modulus Constraint $\beta_\kappa = 0.3$

All the optimized cellular phononic band gap crystals with bulk modulus constraint $\beta_\kappa = 0.3$ and their corresponding band structures for the first eight band gaps of the out-of-plane modes are presented in Figure 7. The first band gap denotes the gap between the first and second band of the diagram, while the second band gap means the gap between the second and the third band of the diagram, and so on. As shown in the band structures, the optimized relative band gap sizes are around 100%. From the perspective of topologies, the complexity is increasing as the target band gap becomes higher. Compared with the optimization topologies of composite phononic crystals, which indicates the heavy and stiff material tends to be isolated by the light and soft material, the solid media also shows the trend to be isolated by air while still connected by thin structures. These slim connections are required so as to support the propagation of the shear components of mechanical waves; otherwise, only longitudinal waves could travel through the totally-isolated solid media. If those slim connections are removed, the remaining topologies are almost the same as those of composite phononic crystals [20].

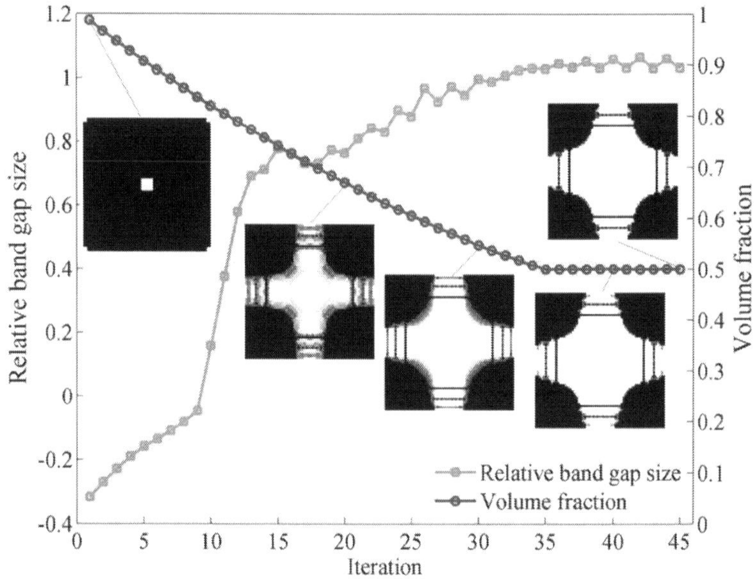

Figure 6. Evolution histories of the relative band gap size, volume fraction and topology in the optimization process with bulk modulus constraint $\beta_\kappa = 0.3$.

Sigmund and Hougaard [41] have investigated the geometric properties of photonic crystals with maximum band gaps and found that the number of inclusions (for transverse electric modes) or subpartitions (for transverse magnetic modes) is interestingly equal to the band number at the lower edge of the band gap. It is also observed that the rule in TE modes of photonic crystals holds for the optimization results for out-of-plane mode of phononic crystals, except that the inclusions are linked to each other by slim connections.

4.1.3. Out-of-Plane Results with Shear Modulus Constraint $\beta_G = 0.3$

Figure 8 presents the optimized phononic band gap crystals with shear modulus constraint $\beta_G = 0.3$ and their corresponding band structures for the first eight band gaps of the out-of-plane modes. The optimized relative band gap sizes with shear modulus constraint are all above 106%, which are larger than the bulk modulus constraint case. In addition, the normalized eigenfrequencies at the lower edge of the gap in the optimized structures are smaller than their counterpart cases with bulk modulus constraints, which indicates results with shear modulus constraints are more favorable for low-frequency-related applications. In terms of optimized topologies, the distributions of solid inclusions within the unit cell are basically akin to these with the bulk modulus constraint case. That is to say the number of the solid parts in the optimized topologies is still equal to the band number. However, the connections are different from those cases with bulk modulus constraint.

4.2. In-Plane Mode

It is more complex to optimize band gap for in plane mode of phononic crystals since both longitudinal and transverse polarizations exist in this mode. This is comparable to the optimization of a complete band gap between TE and TM mode in photonic crystals. Therefore, the resulting optimized structures are more complicated than those for out-of-plane mode.

4.2.1. In-Plane Results with Bulk Modulus Constraint β_K = 0.3

We successfully obtain three optimized topologies among the first six phononic band gap crystals for in plane waves with bulk modulus constraint $\beta_K = 0.3$ as shown in Figure 9. Compared with results for out-of-plane waves, the relative band gap sizes obtained for in-plane waves are smaller while the corresponding topologies are more complicated and have more delicate connections. For example, the optimized structure for fifth band gap has many complicated connections and almost no recognized solid circular or square areas. Moreover, there is no obvious pattern for the distributions of solid material in the optimized structures in this case. Among these results, the third and sixth phononic crystals with maximized band gap are relatively simple and similar to the first and second configurations for out-of-plane waves, respectively.

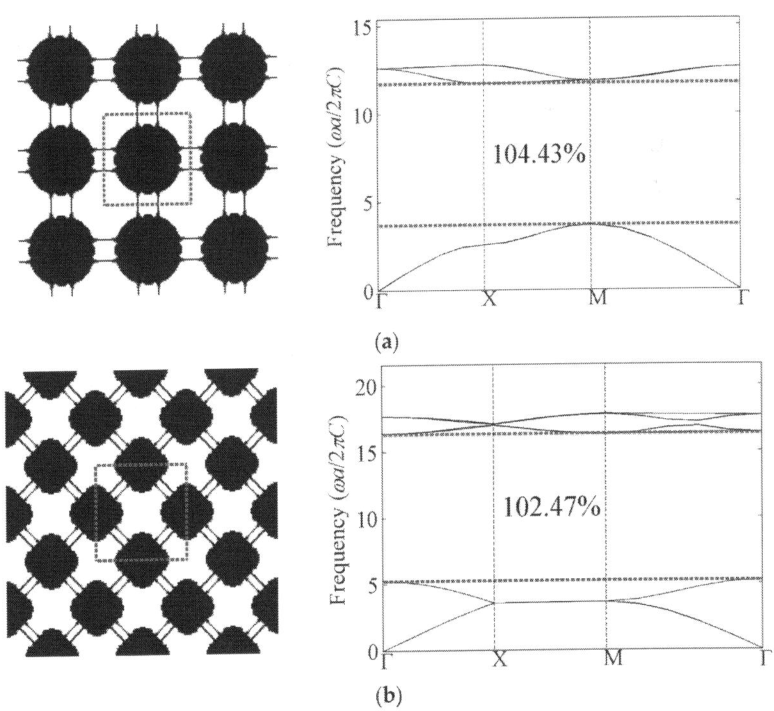

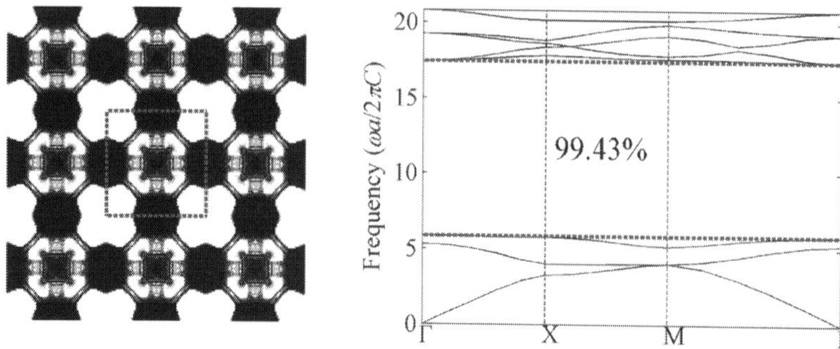

20
15
10
5
0
Frequency (ωa/2πC)
99.43%
Γ
X
M
Γ

(c)

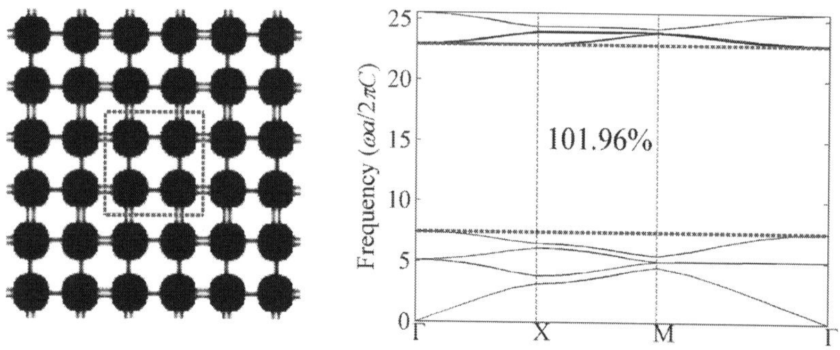

25
20
15
10
5
0
Frequency (ωa/2πC)
101.96%
Γ
X
M
Γ

(d)

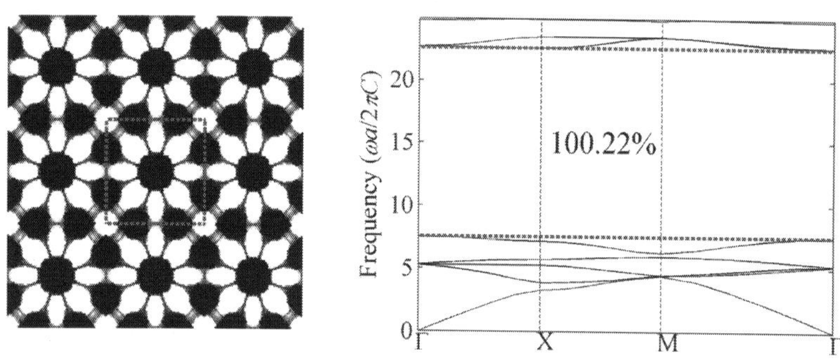

20
15
10
5
0
Frequency (ωa/2πC)
100.22%
Γ
X
M
Γ

(e)

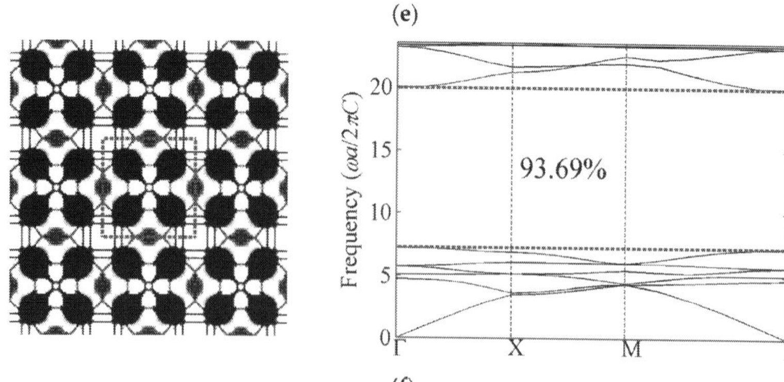

20
15
10
5
0
Frequency (ωa/2πC)
93.69%
Γ
X
M
Γ

(f)

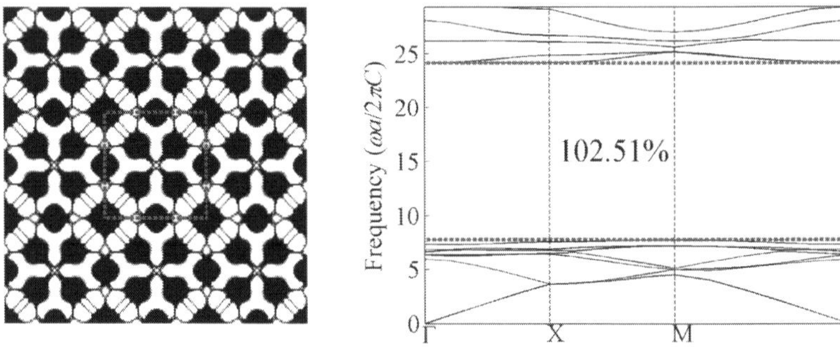

(g)

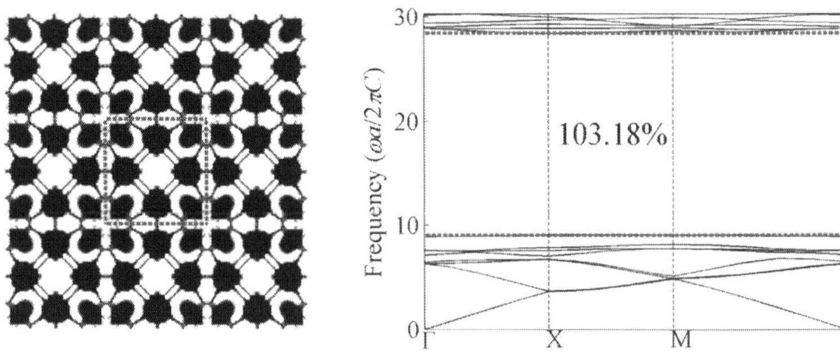

(h)

Figure 7. Optimized topologies and corresponding band structures for out-of-plane mode with bulk modulus constraint. The black and white colors represent silicon and air, respectively. (**a**) The first band gap; (**b**) the second band gap; (**c**) the third band gap; (**d**) the fourth band gap; (**e**) the fifth band gap; (**f**) the sixth band gap; (**g**) the seventh band gap; and (**h**) the eighth band gap.

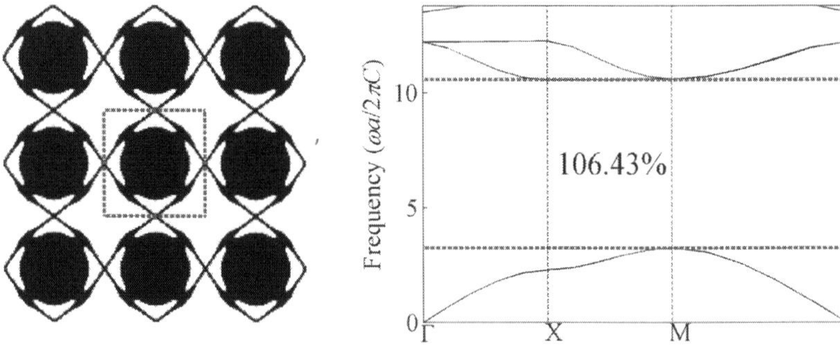

(a)

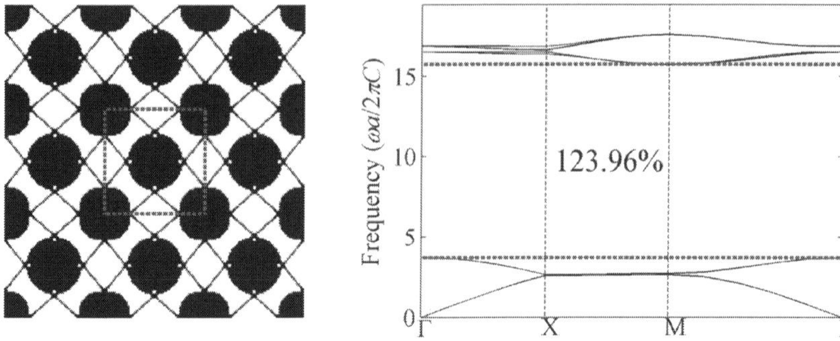
Frequency (ωa/2πC)
0
5
10
15
123.96%
Γ
X
M
Γ

(b)

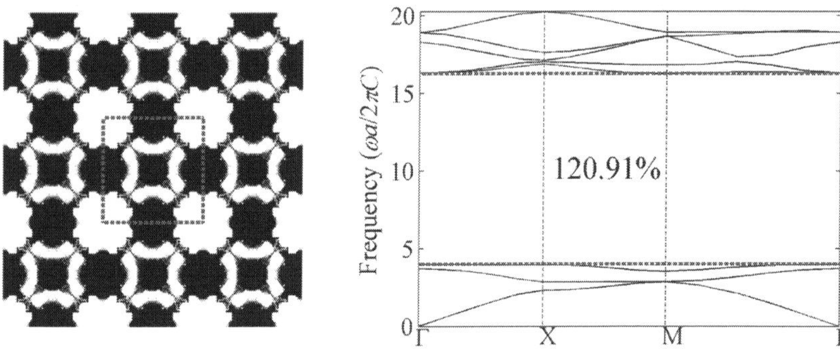
Frequency (ωa/2πC)
0
5
10
15
20
120.91%
Γ
X
M
Γ

(c)

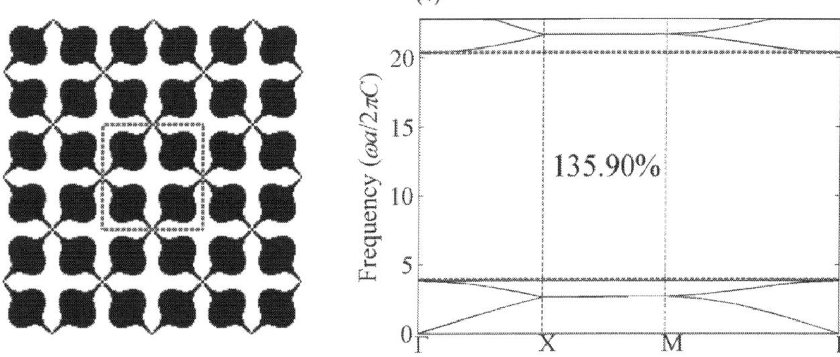
Frequency (ωa/2πC)
0
5
10
15
20
135.90%
Γ
X
M
Γ

(d)

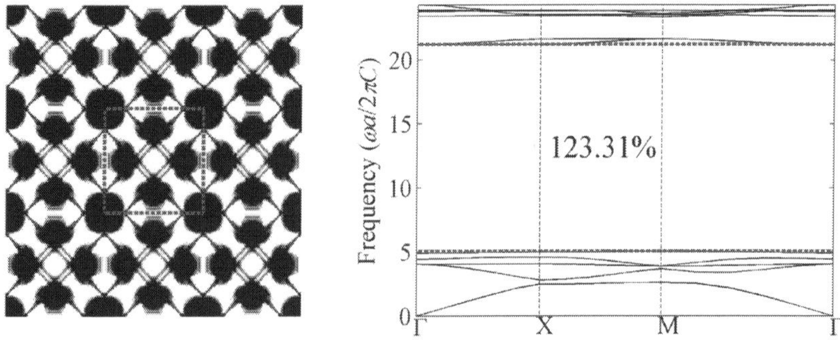
Frequency (ωa/2πC)
0
5
10
15
20
123.31%
Γ
X
M
Γ

(e)

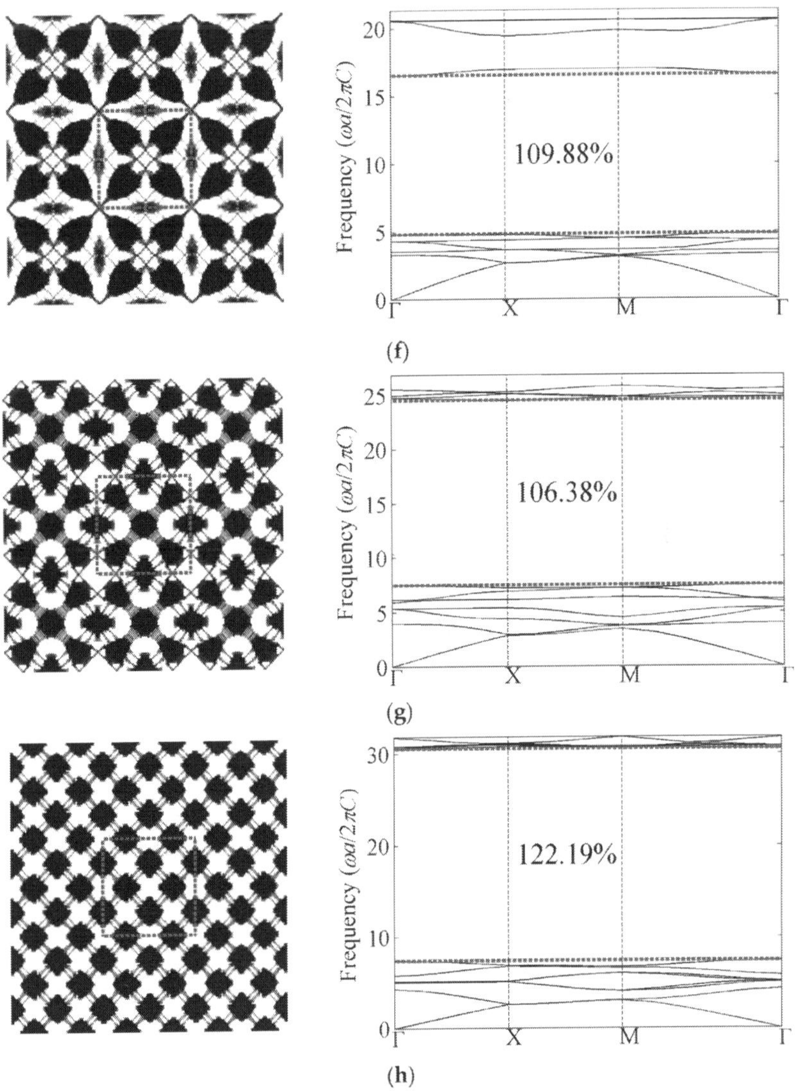

Figure 8. Optimized topologies and corresponding band structures for out-of-plane mode with shear modulus constraint. (**a**) The first band gap; (**b**) the second band gap; (**c**) the third band gap; (**d**) the fourth band gap; (**e**) the fifth band gap; (**f**) the sixth band gap; (**g**) the seventh band gap; and (**h**) the eighth band gap.

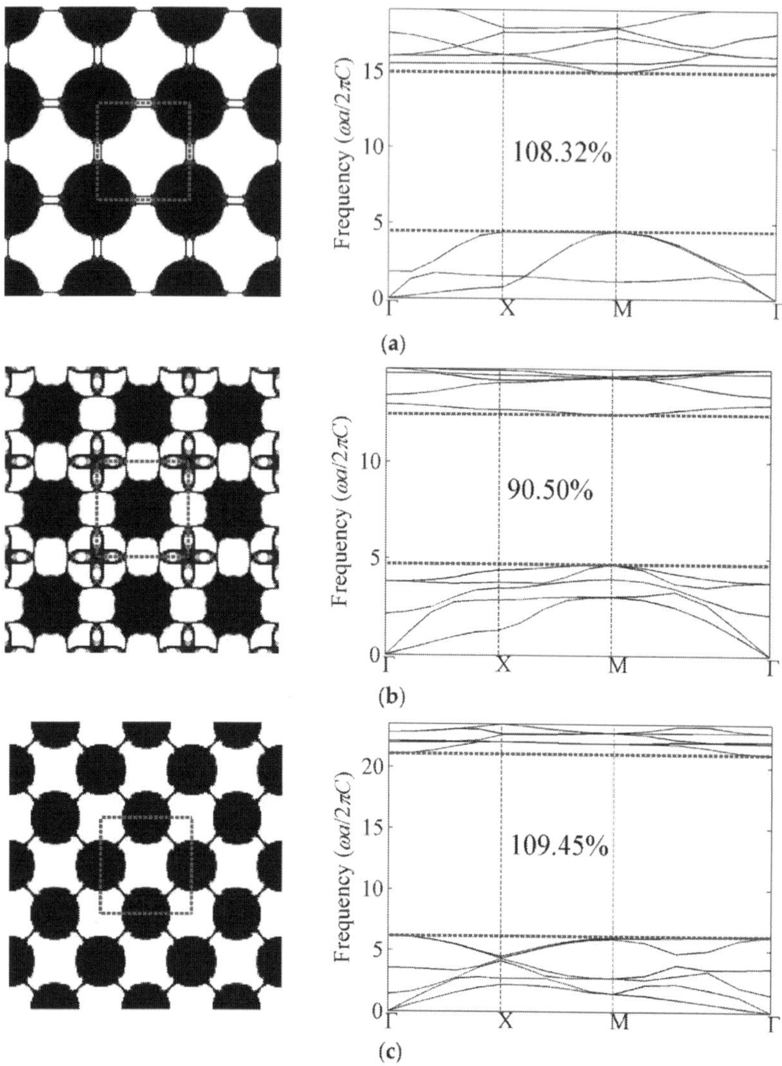

Figure 9. Optimized topologies and corresponding band structures for in-plane mode with bulk modulus constraint. (**a**) The third band gap; (**b**) the fifth band gap; and (**c**) the sixth band gap.

4.2.2. In-Plane Results with Shear Modulus Constraint β_G = 0.3

Optimization results with shear modulus constraint $\beta_G = 0.3$ for in-plane waves are presented in Figure 10. Different from the cases of out-of-plane waves, optimized relative band gap sizes for in plane waves with shear modulus constraint are slightly smaller than that with bulk modulus constraint.

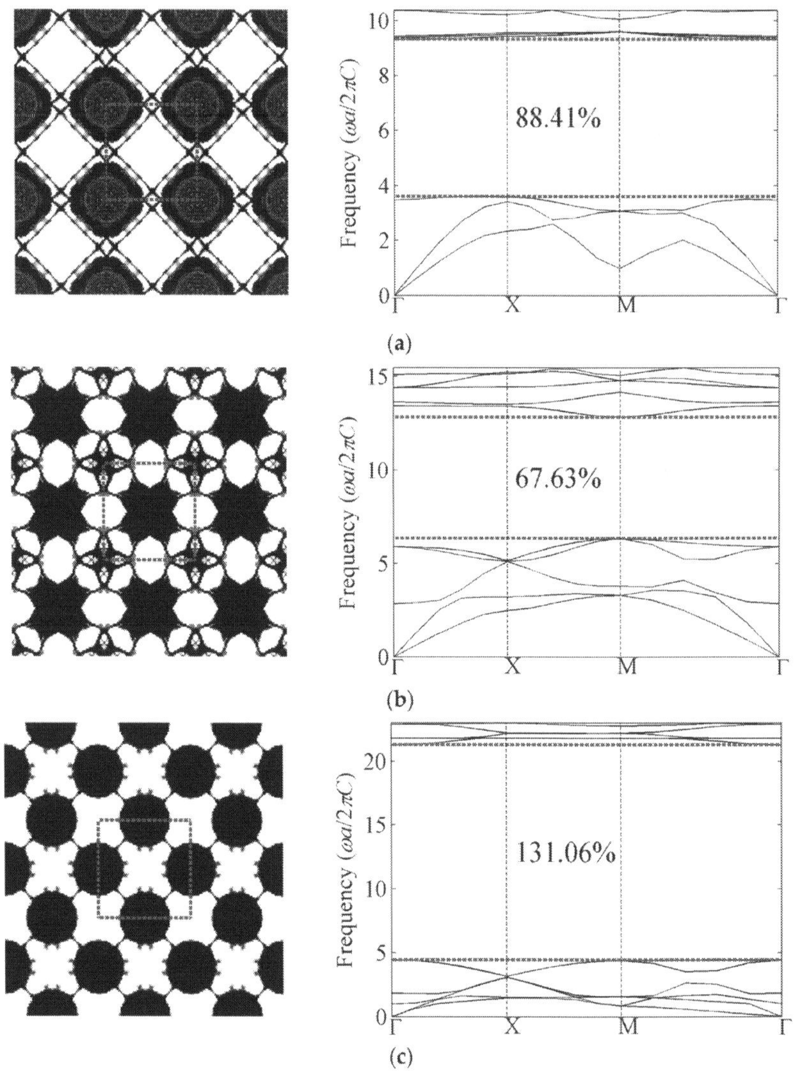

Figure 10. Optimized topologies and corresponding band structures for in-plane mode with shear modulus constraint. (**a**) The third band gap; (**b**) the fifth band gap; and (**c**) the sixth band gap.

Compared with results with bulk modulus constraint, the optimized topologies are similar. It is worthy to mention that a slight variation of connections could lead to dramatic change of the outcome band gap size for instance the sixth band gap. Double-checking the resulting structures with both constraints is recommended in practical application to ensure a larger band gap size.

5. CONCLUSIONS

This paper has systematically investigated the optimized topological design of phononic band gap crystals subject to a volume constraint and a bulk or shear modulus constraint, respectively. The BESO algorithm has been used to seek the optimal distribution of solid material within the square unit cell. The static effective bulk or shear modulus is calculated by a homogenization theory to ensure that the stiffness constraint is satisfied in spite of the variation of the topology in the optimization process. Various optimization results are presented for out-of-plane and in-plane mode and the effects of the stiffness constraint on the topological design of the cellular phononic crystals are also discussed. Numerical results demonstrated that all the optimized structures we obtained for cellular PnCs are similar to these obtained for composite case [20], except that additional slim connections are added in the cellular case, since they are essential to support the propagation of shear wave modes and to satisfy the prescribed bulk or shear modulus constraints. It is also found that the optimal relative band gap sizes dramatically decrease with the increase of stiffness constraint ratio while the slim connections in the corresponding topologies become thicker, which indicates that these connections can be slightly changed based on the manufacture precision on the cost of band gap size. Several optimized designs we obtained have broken the maximum band gap size record in the literature. The main difference between the final results with the bulk modulus constraint and that with the shear modulus constraint relies on the position of the slim connections.

ACKNOWLEDGMENTS

The authors would like to acknowledge the support from Australian Research Council (ARC) Future Fellowship (FT130101094).

AUTHOR CONTRIBUTIONS

The theoretical analysis and numerical simulations in the manuscript were carried out by the PhD candidate Yang Fan Li, and supervised by Xiaodong Huang and Shiwei Zhou.

CONFLICTS OF INTEREST

The authors declare no conflict of interest.

REFERENCES

1. Pennec, Y.; Vasseur, J.O.; Djafari-Rouhani, B.; Dobrzyński, L.; Deymier, P.A. Two-dimensional phononic crystals: Examples and applications. *Surf. Sci. Rep.* **2010**, *65*, 229–291.

2. Sigalas, M.; Kushwaha, M.S.; Economou, E.N.; Kafesaki, M.; Psarobas, I.E.; Steurer, W. Classical vibrational modes in phononic lattices: Theory and experiment. *Z. Kristallogr.* **2005**, *220*, 765–809.

3. Sigalas, M.; Economou, E. Elastic and acoustic wave band structure. *J. Sound. Vib.* **1992**, *158*, 377–382.

4. Kushwaha, M.S.; Halevi, P.; Dobrzynski, L.; Djafari-Rouhani, B. Acoustic band structure of periodic elastic composites. *Phys. Rev. Lett.* **1993**, *71*, 2022–2025.

5. Kushwaha, M.S. Classical band structure of periodic elastic composites. *Int. J. Mod. Phys. B* **1996**, *10*, 977–1094.

6. Sigalas, M.; Economou, E. Attenuation of multiple-scattered sound. *Europhys. Lett.* **1996**, *36*, 241–246.

7. Kafesaki, M.; Economou, E.N. Multiple-scattering theory for three-dimensional periodic acoustic composites. *Phys. Rev. B* **1999**, *601*.

8. Sigalas, M.; Garcıa, N. Theoretical study of three dimensional elastic band gaps with the finite-difference time-domain method. *J. Appl. Phys.* **2000**, *87*, 3122–3125.

9. Tanaka, Y.; Tomoyasu, Y.; Tamura, S.-I. Band structure of acoustic waves in phononic lattices: Two-dimensional composites with large acoustic mismatch. *Phys. Rev. B* **2000**, *62*, 7387–7392.

10. Cao, Y.; Hou, Z.; Liu, Y. Finite difference time domain method for band-structure calculations of two-dimensional phononic crystals. *Solid State Commun.* **2004**, *132*, 539–543.

11. Poulton, C.G.; Movchan, A.B.; McPhedran, R.C.; Nicorovici, N.A.; Antipov, Y.A. Eigenvalue Problems for Doubly Periodic Elastic Structures and Phononic Band Gaps. *Proce. R. Soc.* **2000**, *456*.

12. Axmann, W.; Kuchment, P. An efficient finite element method for computing spectra of photonic and acoustic band-gap materials: I. Scalar case. *J. Comput. Phys.* **1999**, *150*, 468–481.

13. Sigmund, O.; Jensen, J. Systematic design of phononic band-gap materials and structures by topology optimization.*Philos. Trans. R. Soc. Lond.* **2003**, *361*, 1001–1019.

14. Gazonas, G.A.; Weile, D.S.; Wildman, R.; Mohan, A. Genetic algorithm optimization of phononic bandgap structures.*Int. J. Solids Struct.* **2006**, *43*, 5851–5866.

15. Bilal, O.R.; Hussein, M.I. Optimization of phononic crystals for the simultaneous attenuation of out-of-plane and in-plane waves. In Proceedings of the ASME 2011 International Mechanical Engineering Congress and Exposition, Denver, CO, USA, 11–17 November 2011; American Society of Mechanical Engineers: Denver, Colorado, USA, 2011; pp. 969–972.

16. Bilal, O.R.; Hussein, M.I. Ultrawide phononic band gap for combined in-plane and out-of-plane waves. *Phys. Rev. E***2011**, *84*.

17. Dong, H.-W.; Su, X.-X.; Wang, Y.-S. Multi-objective optimization of two-dimensional porous phononic crystals. *J. Phys. D* **2014**, *47*, 155301–155311.

18. Liu, Z.; Wu, B.; He, C. Band-gap optimization of two-dimensional phononic crystals based on genetic algorithm and fpwe. *Wave Random Complex Media* **2014**, *24*, 286–305.

19. Rupp, C.J.; Evgrafov, A.; Maute, K.; Dunn, M.L. Design of phononic materials/structures for surface wave devices using topology optimization. *Struct. Multidiscip. Optim.* **2007**, *34*, 111–121.

20. Dong, H.W.; Su, X.X.; Wang, Y.S.; Zhang, C. Topological optimization of two-dimensional phononic crystals based on the finite element method and genetic algorithm. *Struct. Multidiscip. Optim.* **2014**, *50*, 593–604.

21. Kittel, C.; McEuen, P.; McEuen, P. *Introduction to Solid State Physics*; Wiley: New York, NY, USA, 1976; Volume 8.

22. Kushwaha, M.S.; Halevi, P.; Martínez, G.; Dobrzynski, L.; Djafari-Rouhani, B. Theory of acoustic band structure of periodic elastic composites. *Phys. Rev. B* **1994**, *49*, 2313–2322.

23. Bendsøe, M.P.; Kikuchi, N. Generating optimal topologies in structural design using a homogenization method.*Comput. Methods Appl. Mech. Eng.* **1988**, *71*, 197–224.

24. Sigmund, O. Tailoring materials with prescribed elastic properties. *Mech. Mater.* **1995**, *20*, 351–368.

25. Hashin, Z.; Shtrikman, S. A variational approach to the theory of the elastic behaviour of multiphase materials. *J. Mech. Phys. Solids* **1963**, *11*, 127–140.

26. Xie, Y.; Steven, G.P. A simple evolutionary procedure for structural optimization. *Comput. Struct.* **1993**, *49*, 885–896.

27. Xie, Y.-M.; Steven, G.P. *Basic Evolutionary Structural Optimization*; Springer: London, UK, 1997.

28. Yang, X.; Xei, Y.; Steven, G.; Querin, O. Bidirectional evolutionary method for stiffness optimization. *AIAA J.* **1999**, *37*, 1483–1488.

29. Querin, O.M.; Young, V.; Steven, G.P.; Xie, Y.M. Computational efficiency and validation of bi-directional evolutionary structural optimisation. *Comput. Methods Appl. Mech. Eng.* **2000**, *189*, 559–573.

30. Huang, X.; Xie, Y. Convergent and mesh-independent solutions for the bi-directional evolutionary structural optimization method. *Finite Elem. Anal. Des.* **2007**, *43*, 1039–1049.

31. Huang, X.; Xie, Y. Bi-directional evolutionary topology optimization of continuum structures with one or multiple materials. *Comput. Mech.* **2009**, *43*, 393–401.

32. Huang, X.; Xie, M. *Evolutionary Topology Optimization of Continuum Structures: Methods and Applications*; John Wiley & Sons: Chichester, UK, 2010.

33. Huang, X.; Radman, A.; Xie, Y. Topological design of microstructures of cellular materials for maximum bulk or shear modulus. *Comput. Mater. Sci.* **2011**, *50*, 1861–1870.

34. Huang, X.; Xie, Y.; Jia, B.; Li, Q.; Zhou, S. Evolutionary topology optimization of periodic composites for extremal magnetic permeability and electrical permittivity. *Struct. Multidiscip. Optim.* **2012**, *46*, 385–398.

35. Huang, X.; Zhou, S.; Sun, G.; Li, G.; Xie, Y.M. Topology optimization for microstructures of viscoelastic composite materials. *Comput. Methods Appl. Mech. Eng.* **2015**, *283*, 503–516.

36. Meng, F.; Huang, X.; Jia, B. Bi-directional evolutionary optimization for photonic band gap structures. *J. Comput. Phys.***2015**, *302*, 393–404.

37. Huang, X.; Zuo, Z.; Xie, Y. Evolutionary topological optimization of vibrating continuum structures for natural frequencies. *Comput. Struct.* **2010**, *88*, 357–364.

38. Huang, X.; Xie, Y. Evolutionary topology optimization of continuum structures with an additional displacement constraint. *Struct. Multidiscip. Optim.* **2010**, *40*, 409–416.

39. Komkov, V.; Choi, K.K.; Haug, E.J. *Design Sensitivity Analysis of Structural Systems*; Academic Press: Orlando, FL, USA, 1986; Volume 177.

40. Sigmund, O.; Peterson, J. Numerical instabilities in topology optimization: A survey on procedures dealing with checkerboards, mesh-dependencies and local minima. *Struct. Optim.* **1998**, *16*, 68–75.

41. Sigmund, O.; Hougaard, K. Geometric properties of optimal photonic crystals. *Phys. Rev. Lett.* 2008, *100*.

Index